TEACHER'S

STRATEGIES *for* SUCCESS

MATH Problem Solving

GRADE 8

Strategies for Success: Math Problem Solving, Grade 8, Teacher's Guide OT116TB / 327NATG ISBN-13: 978-1-60161-949-5
Cover Design: Bill Smith Group   **Cover Illustration:** Valeria Petrone/Morgan Gaynin

Triumph Learning® 136 Madison Avenue, 7th Floor, New York, NY 10016 © 2011 Triumph Learning, LLC. Options is an imprint of Triumph Learning®. All rights reserved. No part of this publication may be reproduced in whole or in part, stored in a retrieval system, or transmitted in any form or by any means, electronic, mechanical, photocopying, recording or otherwise, without written permission from the publisher.

Printed in the United States of America.   10 9 8 7 6 5 4 3 2 1

Table of Contents

Program Overview 4

Lesson Plans

Unit 1 Problem Solving Using Real Numbers and Algebraic Thinking 12
    Lesson 1 Work Backward 14
    Lesson 2 Solve a Simpler Problem 18
    Lesson 3 Write an Equation 22
    Lesson 4 Use Logical Reasoning 26
    Unit 1 Review 30

Unit 2 Problem Solving Using Proportional Reasoning 32
    Lesson 5 Make a Table 34
    Lesson 6 Write an Equation 38
    Lesson 7 Work Backward 42
    Lesson 8 Write an Equation 46
    Unit 2 Review 50

Unit 3 Problem Solving Using Algebra 52
    Lesson 9 Make a Graph 54
    Lesson 10 Make a Table 58
    Lesson 11 Guess, Check, and Revise 62
    Lesson 12 Draw a Diagram 66
    Unit 3 Review 70

Unit 4 Problem Solving Using Geometry 72
    Lesson 13 Use Logical Reasoning 74
    Lesson 14 Look for a Pattern 78
    Lesson 15 Draw a Diagram 82
    Lesson 16 Look for a Pattern 86
    Unit 4 Review 90

Unit 5	**Problem Solving Using Measurement**	92
	Lesson 17 Draw a Diagram	94
	Lesson 18 Solve a Simpler Problem	98
	Lesson 19 Guess, Check, and Revise	102
	Lesson 20 Write an Equation	106
	Unit 5 Review	110
Unit 6	**Problem Solving Using Data and Probability**	112
	Lesson 21 Make a Graph	114
	Lesson 22 Look for a Pattern	118
	Lesson 23 Make an Organized List	122
	Lesson 24 Make an Organized List	126
	Unit 6 Review	130

Professional Development Handbook

Problem Solving: Teacher as Coach	133
Reading and Understanding Word Problems	136
Meaningful Practice with Problem Solving	140
Classroom Best Practices	142
Asking Effective Questions	145
Creating a Problem-Solving Environment	146
Using Technology for Teaching and Learning	148
Bibliography	149
Appendix	151

 Plus, **eResources** are available at www.optionspublishing.com.

For each Lesson
- Interactive, whiteboard compatible modelling transparencies
- Homework sheets

For each Unit
- Home-School letters in English and Spanish
- Graphic organizers

See Appendix on page 151 for a complete list of eResources.

3

SUPPORT FOR EVERY TEACHER

STRATEGIES *for* SUCCESS
Math Problem Solving

Finally, a program that implements what research tells us about problem solving!

Focus on Understanding Word Problems

Strategic questioning in each lesson helps students learn how to navigate, read, and understand mathematical text. Understanding a problem involves **paraphrasing**, **visualizing**, **making connections**, and **assuming relationships**.

Model Strategies and Processes

Each lesson provides students with the **guidance** and **tools** they need to fully understand how to approach a problem so they can solve it. The emphasis is on helping students master the problem-solving process.

Build Competence and Confidence

High-interest unit themes engage students from the start. Within each lesson, problems based on the unit theme keep motivation high. **Interactivity** and **scaffolded support** build students' competence and confidence in problem solving.

4   Program Overview

SUCCESS FOR EVERY STUDENT

Program Components

Strategies for Success Math Problem Solving includes Student Worktext, Teacher's Guide, and eResources so you can create a learning environment where all students become successful problem solvers.

Student Worktext

The *Strategies for Success Math Problem Solving* Student Worktext includes:

- a Problem-Solving Toolkit that defines and reinforces the problem-solving strategies and skills students will use throughout the lessons
- ten grade-level problem-solving strategies, each covered two to three times per grade
- 24 lessons providing modeled instruction, guided practice, problem-solving application, and critical thinking questions
- carefully sequenced scaffolding that promotes automaticity of steps used by successful problem solvers

Teacher's Guide

To support instruction, the *Strategies for Success Math Problem Solving Teacher's Guide* offers lesson plans that use strategic questioning to...

- break the problem-solving process into accessible chunks
- help students see relationships between numbers and words
- encourage students to take ownership of the problem-solving process
- promote logical and creative thinking

eResources

The *Strategies for Success Math Problem Solving* eResources include Modeling Transparencies and Reproducible Teacher Resources to support instruction.

- Transparencies are interactive and whiteboard-compatible and can be used to model instruction.
- Additional resources include a self-evaluative problem-solving checklist, vocabulary and problem-solving graphic organizers, family letters, homework worksheets, and mathematical tools such as grid paper and coordinate grids.

Program Overview 5

Instructional Model

Strategies for Success Math Problem Solving follows a consistent instructional model to let you build students' confidence and competence as they learn ways to approach and solve problems.

1. Modeled Instruction

The interactive instruction in each lesson provides a framework that allows students to connect how careful reading and problem-solving strategies work together to reveal the relationships in a problem: **Read the Problem, Search for Information, Decide What to Do, Use Your Ideas, Review Your Work.** Guiding questions model logical and creative thinking, while hints and visual supports help deepen conceptual understanding.

2. Guided Practice

With scaffolded practice, the problem-solving strategy is applied to similar problem situations that take students further in their thinking and reasoning. The hints and guided questioning that strengthen students' ability to self-monitor are gradually removed as students gain understanding, skills, and confidence.

3. Independent Practice

Students demonstrate their understanding by working on problems without scaffolding. The critical thinking questions that conclude each problem provide additional cognitive support for making the complex connections needed in active problem solving. At the end of every lesson, students modify or create a problem as a way to consolidate their learning.

Additional Support

Differentiated Instruction *Strategies for Success Math Problem Solving* addresses the needs of learners at all levels of proficiency by providing language support for English language learners, activities that address the concepts most likely to be stumbling blocks for struggling students, and techniques for teachers to modify problems for students who are competent problem solvers.

Progress Monitoring All lesson problems and the Create activity can be used as informal assessment to evaluate students' facility with the problem-solving process. The unit review can be used as formal assessment to test students' understanding of the problem-solving strategies they have learned within each unit.

Home/School Connection Each **Family Letter** explains an aspect of good problem solving and includes brief activities that adults can work on with their students. **Strategy Focus Worksheets** mirror the lesson in the student worktext and provide scaffolding to help students and their family members understand how to use the focus strategy to solve each problem.

6   Program Overview

Student Worktext

Strategies for Success Math Problem Solving helps your students take ownership of the problem-solving process.

- **Unit Opener** establishes the motivating **theme** of the unit, introduces mathematical **vocabulary**, and uses vocabulary activities and **graphic organizers** to further deepen understanding of mathematical language.

- **Lessons** are designed for engagement and target grade-level proficiency. Each six-page lesson focuses on a single problem-solving strategy and includes **modeled instruction** that engages students in the thinking process, **guided practice** that lets students build their own understanding of how to solve problems, and **independent practice** that creates a learning environment where students can figure out approaches on their own.

- **Unit Review** serves as both **practice** and **assessment**. In the unit review, students show their work and record their answer and the strategy they used to solve each problem. In two problems, students explain their reasoning. The review includes a writing prompt and a small group, **collaborative decision-making activity** to provide further opportunities for students to demonstrate understanding of the problem-solving strategies in the unit.

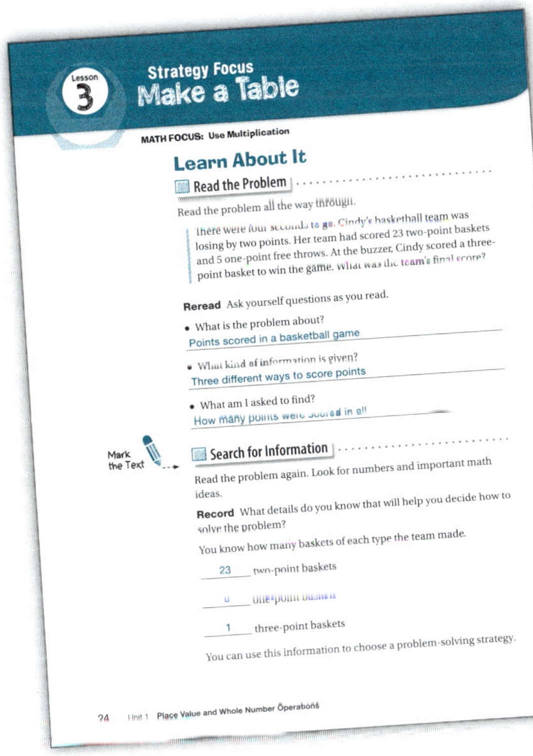

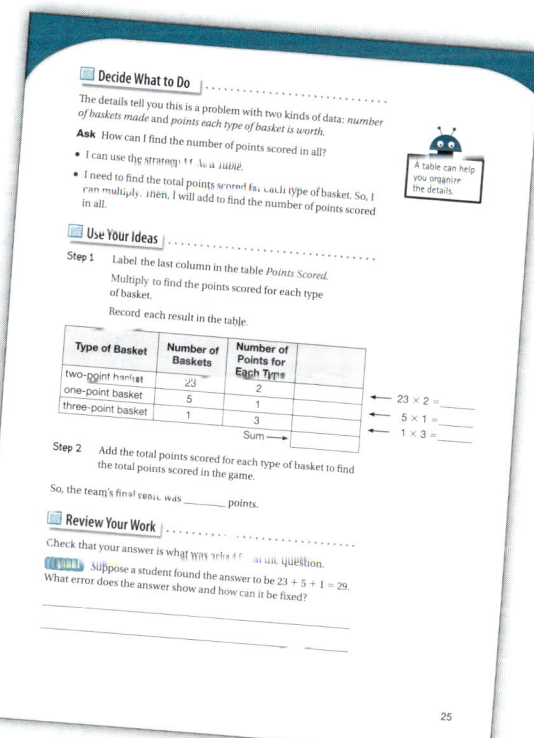

Turn the page to see the support materials for this lesson. ➤

Program Overview   7

Teacher's Guide

Strategies for Success Math Problem Solving gives you the tools you need for teaching problem-solving strategies and processes and provides you with support every step of the way.

Teacher eResources

Outcome
Instructional purpose of lesson

Lesson Overview

Lesson Materials: calculator, ruler

Skills to Know	Outcome	Math Vocabulary	eResources www.optionspublishing.com
• Subtract 3-digit numbers from 4-digit numbers • Multiply up to 3-digit numbers by 2-digit numbers	Students will recognize that tables are an efficient way to organize information in problems with different kinds of data.	Transformation	• Problem-Solving Checklist, also available in the student worktext, page 7 • Interactive Whiteboard Transparency 3 • Homework, Unit 1 Lesson 3 • Know-Find-Use Graphic Organizers (optional)

Modeled Instruction
Helps students monitor comprehension as they respond to questions that clarify understanding.

Modeled Instruction
Learn About It

Ask questions to confirm students' comprehension of the problem's context.
- How many ways are there to score points in a basketball game?
- How can you ask the question in the problem another way?

As students read the problem again, pose questions to help identify important information.
- Are there any numbers in the problem you do not need? How did you decide?
- How can you tell the number of three-point baskets the team scored?

Comprehensive Instruction in every lesson
includes modeling the problem-solving process, addresses the mathematics within each problem, and points out common student pitfalls—lesson by lesson.

 Think Aloud You may wish to use this to demonstrate how to read a problem for different purposes.

Read the problem aloud.
I know the problem is about basketball, and there are several numbers. Most of the numbers are about scoring points, but there is also a number about time.
I am going to reread the problem to decide which numbers I need to solve it. If I find what the problem is asking I will be able to tell what numbers I need.
I see that the problem asks me to find how many points Cindy scored to win the game. Wait—I read too quickly. The problem asks how many points Cindy's team scored, not how many points Cindy scored.

Think Aloud
Models Reread, Represent, or Use Graphic Organizer.

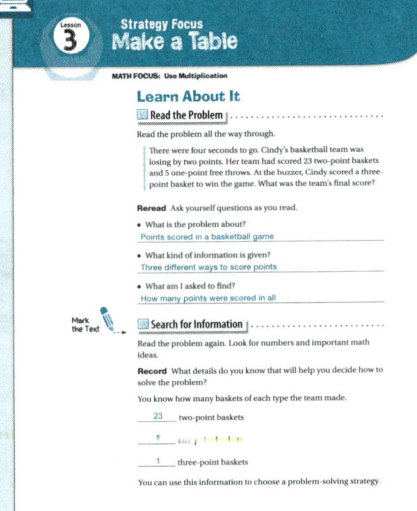

I think I should start by finding how many different ways Cindy's team scored points. I do not need the detail about four seconds. That is not about scoring points. I am going to cross that out. I do not need the detail about how far behind her team was, either. I will cross that out too.

6 Unit 1

8 Program Overview

Differentiated Instruction provides customized options for meeting the individualized needs of all your students.

Professional Development incorporates methods and ideas from the latest research on a variety of topics key to teaching problem solving, including how to read word problems effectively, writing and communicating about math, how to determine what makes a good problem, and more.

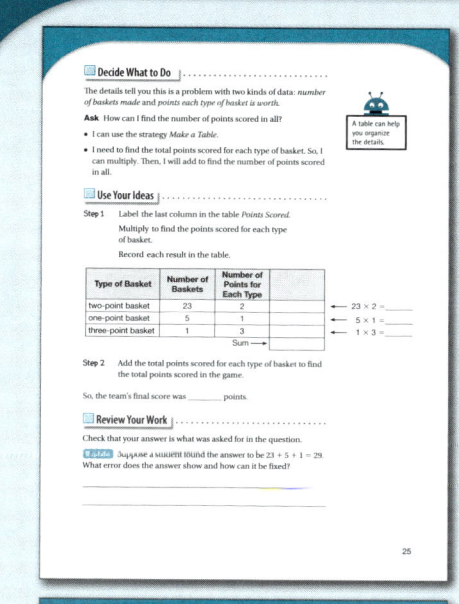

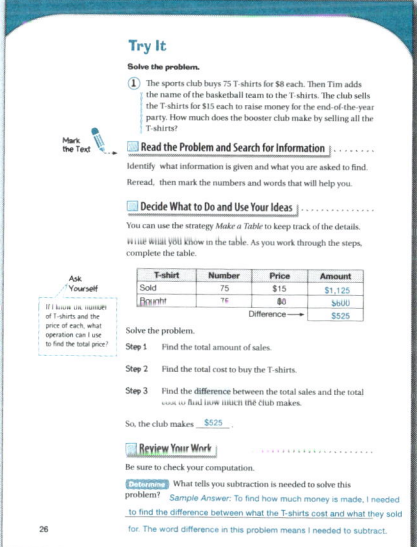

Modeled Instruction (continued)

Help students make a connection between what they know and what they need to find out.

- How will a table help you organize the details?
- How can you tell there is more than one step for solving this problem?

Pose questions that give meaning to each step in the process.

- What information does 23 × 2 give? 5 × 1? 1 × 3?
- After you have multiplied to find the points scored for each type of basket, what do you need to do next?

Emphasize the importance of answering the right question by posing other questions about the problem.

- How would the answer change if Cindy did not score a three-point basket at the end of the game?

HOTS Explain Students' responses should demonstrate they recognize and can correct commons errors made when solving problems similar to this.

HOTS Questioning
Strengthens students' cognitive processing through a higher order thinking skill prompt.

Guided Practice
Try It

(1) Prompt students to consider the relationships among the problem parts.

Discuss the context of the problem as well as the key data for solving it.

- Why is the price of the T-shirts the club sells different from the price they bought them for?

Have students provide the number sentences they used to solve the problem.

- Does the money collected selling the T-shirts tell you how much money the club made? Why or why not?

Some students may recognize that they can first find 15 − 8 (the profit for one T-shirt), and multiply 75 by the result to find how much the club made.

HOTS Determine Students' explanations should demonstrate how they determined the operation needed for the problem solution.

Guided Practice
Allows students to consolidate and apply what they have learned as well as extend their thinking.

Lesson 3 7

Teacher eResources

Includes interactive, whiteboard-ready Modeling Transparencies for each lesson, a student problem-solving checklist, family letters in English and Spanish, homework strategy focus worksheets, vocabulary and problem-solving graphic organizers, and additional reproducible teacher resources.

Program Overview 9

Implementation Guide

Strategies for Success Math Problem Solving is flexible. You can teach problem solving in a whole-class setting. Or you can use the program for small-group instruction, after school, during summer school, or in other implementations outside the classroom.

Enhance Your Core Program

Twenty-four lessons provide instruction for a weekly class in problem solving. You can substitute the **Strategies for Success Math Problem Solving** lessons for the strategy lessons in your core program.

Build Skills for Students Who Need Extra Support

Systematic, scaffolded instruction is the perfect model for after school, and summer school. Students are provided a safety net that helps develop organizational skills, while reinforcing mathematical ideas.

A variety of visual devices and language support ensure that English language learners understand academic vocabulary. Oral and written language skills are continually developed and reinforced.

Prepare for the Common Core State Standards

Problem solving is the heart of mathematics. A sound understanding of problem-solving strategies helps students build competence and fluency in all aspects of learning mathematics. This underlying principle is the focus of the new approach to mathematics curriculum designs outlined in the Common Core State Standards.

The carefully scaffolded lessons in **Strategies for Success Math Problem Solving** develop key strategic skills that help students understand how to solve different kinds of problems, such as looking for patterns, organizing information, and representing situations.

Pacing and Classroom Management

The lessons in *Strategies for Success Math Problem Solving* can conform to the scheduling needs of any classroom. Teachers may follow the suggested plan for presenting each lesson, or modify the plan to meet the needs of their students.

Whole Class

Instructional Uses	Component	Pacing
Replace or supplement problem-solving strand in core curriculum	*Strategies for Success Math Problem Solving* Student Worktext	24 weeks 1 lesson per week (15–20 minutes per day or two 30–45 minute sessions)
	Strategy Focus Worksheet	24 weeks 1 lesson per week

Small Group

Instructional Uses	Component	Pacing
After-school program for students needing extra help **OR** Vocabulary support for English language learners	*Strategies for Success Math Problem Solving* Student Worktext	24 sessions 1 lesson per session ($1-1\frac{1}{2}$ hours)
	Strategy Focus Worksheet	24 sessions 1 lesson per session

Summer School

Instructional Uses	Component	Pacing
Problem-solving program to meet needs of special groups (can be used either with mathematically able or mathematically struggling students, depending on grade level chosen)	*Strategies for Success Math Problem Solving* Student Worktext	6 weeks 1 lesson per day ($1-1\frac{1}{2}$ hours)

Strategies for Success Math Problem Solving and Your Core Math Program

Basal programs typically rely on a heuristic—Understand, Plan, Solve, Look Back—to structure the problem-solving process, but then provide little in the way of instruction about how to apply the process. The instruction and scaffolding of each student lesson in *Strategies for Success Math Problem Solving* lead students to internalize a process where they read and reread a problem, extract important words and numbers, visualize the action in a mathematical situation, identify the question being asked, organize information to see relationships, decide what to do to solve the problem, complete a solution plan, and review their work to consolidate what they have learned.

UNIT 1: Problem Solving Using Real Numbers and Algebraic Thinking

CCSS 8.EE Expressions and Equations

Unit Overview

Lesson	Problem-Solving Strategy	Math Focus
1	Work Backward	Rational Numbers
2	Solve a Simpler Problem	Exponents and Scientific Notation
3	Write an Equation	Operations with Algebraic Expressions
4	Use Logical Reasoning	Solve Equations and Inequalities

Promoting Critical Thinking

Higher order thinking questions occur throughout the unit and are identified by this icon: 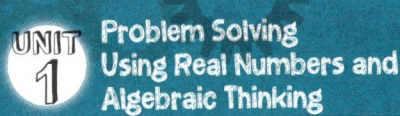. These questions progress through the cognitive processes of remembering, understanding, applying, analyzing, evaluating, and creating to engage students at all levels of critical thinking.

UNIT 1: Problem Solving Using Real Numbers and Algebraic Thinking

Unit Theme: Challenges

Life is full of challenges. There are worldwide challenges, such as curing diseases. There are scientific challenges, such as exploring outer space. There are also personal challenges, such as achieving a goal you have set for yourself. In this unit, you will find that math is often used when facing all kinds of challenges.

Math to Know

In this unit, you will apply these math skills:
- Perform basic operations with rational numbers
- Compute with exponents and write numbers using scientific notation
- Use algebraic expressions, equations, and inequalities to solve problems

Problem-Solving Strategies
- Work Backward
- Solve a Simpler Problem
- Write an Equation
- Use Logical Reasoning

Link to the Theme

Write another paragraph about the park's restoration. Include some of the words and numbers from the table.

The city does not have the money to make improvements to a local park, so Rory's class has volunteered to make them for free. The class wants to finish within two weeks. The students make a table to organize some of their tasks.

Task	Students Needed	Time for Task
Pick up trash	4	2 days
Paint benches	6	4 days
Plant garden	5	5 days

Students' paragraphs will vary, but should include
some words and numbers from the table.

Use Math Language

Review Vocabulary

The list below shows vocabulary terms in this unit. Knowing the meaning of these terms will help you understand the problems.

base exponent inequality scientific notation word equation
equation expression inverse operation variable

Vocabulary Activity Modifiers

A descriptive word placed in front of another word can indicate a specific meaning. When learning new math vocabulary, pay attention to each word in the term. Use terms from the list above to complete the following sentences.

1. The __inverse operation__ of multiplication is division because one is used to undo the other.
2. 8,200 can be expressed in __scientific notation__ as 8.2×10^3.
3. A __word equation__ shows the relationship in words between the important quantities in a word problem.

Graphic Organizer Word Circle
Complete the graphic organizer.
- Cross out the word that does not belong.
- Replace it with a word from the vocabulary list that does belong.
- Write an example for each word.

Sample Answers:

variable	expression
In $x - 5 = 12$, x is the variable.	$6x$
inequality	~~exponent~~ equation
$1 > -5$	$4x = 20$

22   23

Link to the Theme Challenges

Ask students to read the direction line and story starter. If students are having trouble getting started, ask questions such as, *How many students will work to improve the local park? How long will each student work? What other tasks might students do?*

12

Unit 1 Differentiated Instruction

Extra Support

Some students may need to review scientific notation before they solve problems involving numbers expressed in scientific notation.

Proper Form of Scientific Notation Write 16.3×10^3 on the board. Ask, *Is this number written in the proper form for scientific notation? Why or why not?* Then model, step by step, how to rewrite it as 1.63×10^4. Repeat the process with 0.61×10^7. Then ask students to write similar numbers in scientific notation.

Multiplication with Scientific Notation Demonstrate the reason for adding exponents when multiplying numbers written in scientific notation.

$(2.3 \times 10^3) \times (3.1 \times 10^2)$
$= (2.3 \times 3.1) \times (10^3 \times 10^2)$
$= 7.13 \times (10 \times 10 \times 10) \times (10 \times 10)$
$= 7.13 \times 10^5$

Then demonstrate division with numbers written in scientific notation.

$$\frac{4.8 \times 10^4}{3.2 \times 10^2} = \frac{4.8 \times (10 \times 10 \times 10 \times 10)}{3.2 \times (10 \times 10)}$$
$$= 1.5 \times 10^2$$

Challenge Early Finishers

Students may benefit by using the same information to answer questions involving $\leq$, $=$, and $\geq$. Suggest a situation such as this one.

Dolores has 30 packages each of two types of trail mix. She sells the fruit and nut mix for $1.50 per package. She sells the chocolate and nut mix for $2.00 per package.

Then have students answer questions such as these.

- *If she collected $58, what are three combinations of packages she could have sold?*
- *Suppose she collected at least $75. What are three combinations of packages she could have sold?*
- *Roger cannot spend more than $25. What are three combinations of packages he could buy?*

English Language Learners

Reading Comprehension

Per Pound, Per Hour, Per Month Write the following sentences from the unit on the board:

- The gear on Mr. Abrams's sled is worth $75 per pound.
- The family from Alta traveled upstream at 4 miles per hour.
- A teak plantation can produce about 20.4 cubic meters of wood per month.

Explain that one meaning of the word *per* is "for each." Point to the first sentence and say, *So another way to say this sentence is "The gear on Mr. Abrams's sled is worth $75 for each pound."* Have volunteers rephrase the remaining sentences in a similar manner. Then have pairs ask and answer questions that include *per*, such as, *How many books do you read per year?*

Vocabulary

Prefixes Write *undo* on the board and remind students that when they work backward to solve a problem, they undo the actions described in the problem. Then circle *un-* and explain that it is a prefix, which is a word part added at the beginning of a word that changes the meaning of the word. Explain that the prefix *un-* can mean to reverse something. So when you *undo* a set of actions, you reverse what has been done. Demonstrate this meaning of the prefix *un-* by tying and untying your shoelaces, or locking and unlocking a door or window.

Writing

Word Study To help students better understand the problems in this unit, have them scan the unit to select three unfamiliar words or words they would like to know more about. They should look up each word in a dictionary or other reference and then write a sentence using that word. Words from the unit students may identify include *capacity, circumference, satellite, carabiner, canoe, tropical,* or *canal*. Have volunteers read their sentences aloud.

Lesson 1: Strategy Focus — Work Backward

Lesson Overview

Lesson Materials: calculator

Skills to Know	Outcome	Math Vocabulary	eResources www.optionspublishing.com
• Operations with whole numbers • Operations with fractions • Operations with decimals	Students will recognize that working backward is an efficient way to use a final amount to find an initial amount.	equation, expression, inverse operation	• Interactive Whiteboard Transparency 1 • Homework, Unit 1 Lesson 1 • Problem-Solving Checklist, also available in the student worktext, page 7

Modeled Instruction

Learn

Ask questions about the problem's context to clarify students' comprehension of what the problem is about.

- *Why is the community building wildlife tunnels?*
- *Which sentence tells you that the tunnels are part of a larger project?*

As students read the problem again, ask questions to help them focus on the details needed to solve the problem.

- *What does the number 5,450,000 represent?*
- *How does the total cost compare to the cost of the four tunnels?*

 Reread You may wish to use this Think Aloud to demonstrate determining how to use the given information to decide which problem-solving strategy to use.

I know this problem is about a construction project. The problem gives the total cost of the project. It also tells me how the cost of building 4 tunnels compares to the total cost. I need to figure out how much each tunnel costs.

I wish the problem gave the cost of one tunnel and asked for the total cost. I could just multiply by 4 for four tunnels. Then I would multiply by $12\frac{1}{2}$ and add $75,000. That would give me the total cost. But I already know the total cost. I need to find the beginning number. The problem is backward. Maybe I could work backward.

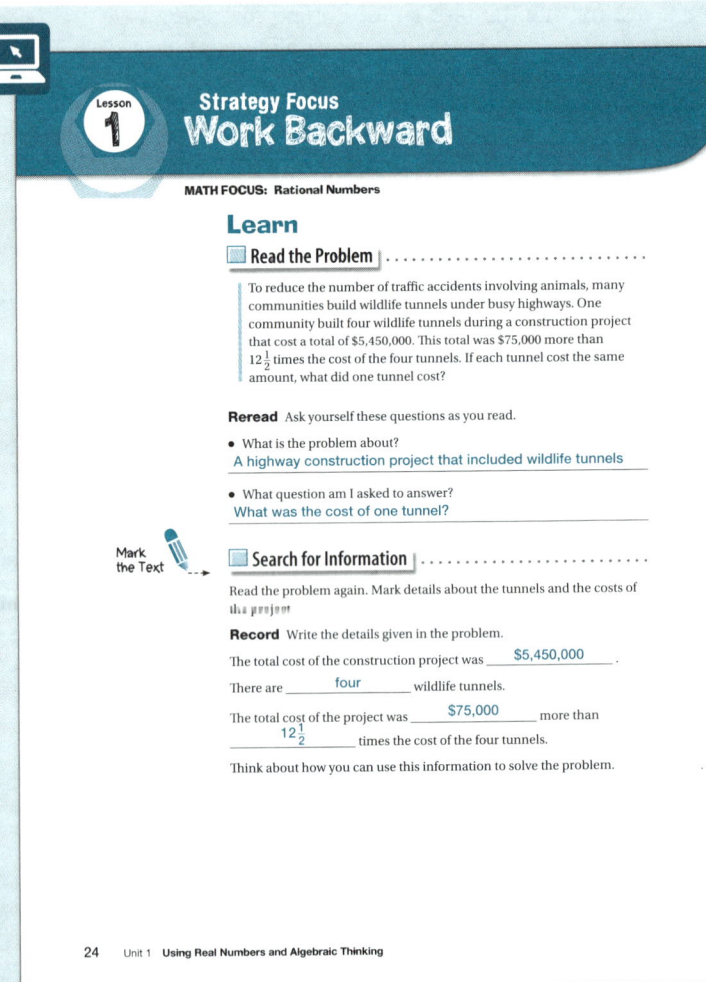

14   Unit 1

Decide What to Do

You know the total cost of the construction project and the number of tunnels. You also know the total cost of all four tunnels expressed in terms of the total cost of the project.

Ask How can I find the cost of one tunnel?

- I can use the strategy *Work Backward*. I know the total cost of the project. I can use that amount to find the cost of four tunnels.
- I know how the costs of the project are related, so I can use inverse operations to undo the relationships and find the cost of one tunnel.

Use Your Ideas

Step 1 Writing an equation using words and numbers can help you see how the numbers are related.

Cost of 1 tunnel $\times 4 \times 12\frac{1}{2} + \$75{,}000 = \$5{,}450{,}000$

Step 2 Draw a diagram to show the operations going forward and how they will be reversed.

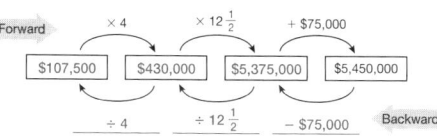

When you work backward, remember to use inverse operations.

Step 3 Work backward to find the cost of one tunnel.

One wildlife tunnel cost $\underline{\$107{,}500}$.

Review Your Work

Check to be sure that for every operation going forward, you used the inverse operation going backward.

Describe How did you determine the inverse operations to use in order to work backward?

Sample Answer: Using inverse operations means undoing an operation. So I subtracted to undo the addition and I divided to undo the multiplication.

25

Modeled Instruction (continued)

Help students connect the facts they know with a strategy they can use to solve the problem.

- *If you knew the cost of one tunnel, how could you find the total cost of the project?*
- *How could you reverse that process to find the cost of one tunnel?*

Ask questions that guide students to consider each step in the solution process.

- *How does the diagram help you understand how to work backward?*

Emphasize the importance of confirming that the correct operations were used.

HOTS Describe Responses should show that students understand inverse operations.

Try

Solve the problem.

1. Elevators move up and down with the help of a heavy weight called a counterweight. Engineers find the weight of a counterweight by subtracting 250 pounds from the maximum capacity of the elevator, and then taking half of the difference. What is the maximum capacity of an elevator that uses a 500-pound counterweight?

Mark the Text

Read the Problem and Search for Information

Restate the problem in your own words. Think about how the weight of a counterweight is related to the maximum capacity.

Decide What to Do and Use Your Ideas

You know how to find the weight of a counterweight when given the maximum elevator capacity. You can work backward because the weight of a counterweight is given.

Ask Yourself What words can I use to describe the problem?

Step 1 Write an equation to represent the problem situation.

Counterweight = (Maximum capacity − $\underline{250}$ pounds) × $\underline{\frac{1}{2}}$

Step 2 Draw a diagram to help you work backward.

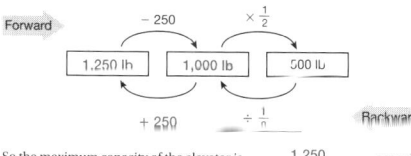

So the maximum capacity of the elevator is $\underline{1{,}250}$ pounds.

Review Your Work

Check that your calculations work going forward and backward.

Compare Arlo says that in order to get the number in the middle box of the diagram, he multiplied by 2 instead of dividing by $\frac{1}{2}$. Explain why Arlo's method works.

Sample Answer: When dividing by a fraction, you multiply by the reciprocal of the divisor. The reciprocal of $\frac{1}{2}$ is $\frac{2}{1}$, or 2, so dividing by $\frac{1}{2}$ and multiplying by 2 have the same result.

26 Unit 1 Using Real Numbers and Algebraic Thinking

Guided Practice

Try

1. Help students connect the information in the problem with the operations needed to solve it.

Ask students to state the problem in their own words.

- *What does* maximum capacity *mean?*
- *If the engineers know the maximum capacity, how do they find the weight of the counterweight?*

Help students focus on the inverse operations needed to solve the problem.

- *What is the opposite of finding half of a number?*
- *What is the opposite of finding a difference?*

Make sure students check their answers by working forward through the operations in the diagram.

HOTS Compare Students' responses should demonstrate that they understand that dividing by a fraction is the same as multiplying by its reciprocal.

Lesson 1   15

Scaffolded Practice

Apply

2 Guide students to understand why and how they would work backward to solve the problem.

- *If you knew the length of the Panama Canal, how could you find the length of the Erie Canal?*

HOTS **Clarify** Students should provide an expression equivalent to $7p + 13$, where p represents the length of the Panama Canal in miles.

3 Make sure students understand the situation described in the problem.

- *What does the first sentence in the problem mean?*
- *How does the amount the locks on the Erie Canal raise barges compare to the amount the locks on the Panama Canal raise ships?*

HOTS **Determine** Explanations should show that students understand that the phrase *$7\frac{1}{2}$ times as high* shows that the locks on the Erie Canal raise barges higher.

4 Help students determine how to work backward to solve this problem.

- *Does the order in which you perform the operations matter when solving this problem? Why or why not?*

HOTS **Sequence** Students' responses should note that they needed to find the difference between the amount required and the amount still needed.

5 Ask questions to help students think about working backward by breaking the problem into steps.

- *How can you find the length of the lock?*
- *If you know the length of the lock, how can you find its width?*

HOTS **Explain** Students' explanations should include a description of how to work from the beginning to check that they get the same final number.

Apply

Solve the problems.

2 The Erie Canal was about 13 miles longer than seven times the length of today's Panama Canal. The Erie Canal was 363 miles long. About how long is the Panama Canal?

Ask Yourself How can I organize the information given in this problem?

Hint Start with the length you know to solve the problem.

Forward: 50 mi → $\times 7$ → 350 mi → $+ 13$ → 363 mi

Backward: 363 mi → $- 13$ → 350 mi → $\div 7$ → 50 mi

Answer The Panama Canal is about 50 miles long.

Clarify What expression can you write to describe the length of the Erie Canal in terms of the length of the Panama Canal?

Sample Answer: If P represents the length of the Panama Canal in miles, I could express the length of the Erie Canal as $7P + 13$.

3 The locks on the Panama Canal raise ships from sea level to the continental divide. The locks on the Erie Canal raised barges 555 feet. That is $82\frac{1}{2}$ feet less than $7\frac{1}{2}$ times as high as the Panama Canal raises ships. What is the altitude of the continental divide where the Panama Canal crosses it?

Hint You can write an equation to describe the situation.

Ask Yourself What expression describes $82\frac{1}{2}$ feet less than $7\frac{1}{2}$ times some number?

Forward: 85 ft → $\times 7\frac{1}{2}$ → $637\frac{1}{2}$ ft → $- 82\frac{1}{2}$ → 555 ft

Backward: 555 ft → $+ 82\frac{1}{2}$ → $637\frac{1}{2}$ ft → $\div 7\frac{1}{2}$ → 85 ft

Answer The altitude of the continental divide where the Panama Canal crosses it is 85 feet.

Determine What were you able to infer from your first reading of this problem about the relative heights of the two canals?

Sample Answer: From the phrase *$7\frac{1}{2}$ times as high*, I knew that the Erie Canal raised ships much higher than the Panama Canal.

4 R-value measures resistance to heat flow in home insulation. Leah's home already has some insulation with an R-value of 3.70 per inch. To get an R-value of 24.05, it needs 4.5 more inches of insulation. How many inches of insulation does the home have?

Hint The second sentence means that each inch of additional insulation adds 3.70 to the total R-value.

Ask Yourself What equation can I write to represent find the total R-value?

Forward: 2 → $+ 4.5$ → 6.50 → $\times 3.70$ → 24.05

Backward: 24.05 → $\div 3.70$ → 6.50 → $- 4.5$ → 2

Answer The home has 2 inches of insulation.

Sequence After you found the total inches of insulation needed, how did you find the amount of insulation already in the home?

Sample Answer: I subtracted the amount needed from the total number of inches required for an R-value of 24.05.

5 The locks along the Panama Canal are rectangular. The length of each lock is about 3.5 meters more than 9 times its width. The largest ship the locks can carry is 294 meters long. That is 11 meters less than the length of a lock. About how wide is a Panama Canal lock?

Hint First, find the length of the lock.

Ask Yourself In what order should I perform the operations to find the width?

Forward: 33.5 m → $\times 9$ → 301.5 m → $+ 3.5$ → 305 m → $- 11$ → 294 m

Backward: 294 m → $+ 11$ → 305 m → $- 3.5$ → 301.5 m → $\div 9$ → 33.5 m

Answer A Panama Canal lock is about 33.5 meters wide.

Explain How can you use your diagram to check your answer?

Sample Answer: If I start with my answer and do the calculations in each forward step, I should get to 294 meters at the end.

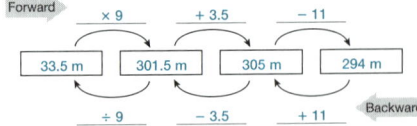

Practice

Solve the problems. Show your work.

6 In 1837, about 500,000 bushels of wheat were shipped using the Erie Canal. That was 9,600 bushels less than 140 times the amount shipped in 1829. How many bushels of wheat were shipped on the Erie Canal in 1829?

Some students will use the lesson strategy; however, other strategies may be used. Accept all reasonable work leading to the correct answer.

Answer About 3,640 bushels of wheat were shipped on the Erie Canal in 1829.

 Identify Which problem in this lesson is most like Problem 6? Explain.

Sample Answer: Problem 3, because both problems require two steps and involve addition and division when working backward.

7 A *span* is the distance between two bridge supports. The span of an arch bridge from Island A to Island B is 5 times the length of the span of a beam bridge between the two islands. The distance from Island A to Island B is 7,956 feet. That is 44 feet less than 8 times the span of the arch bridge. What is the span of a beam bridge?

Some students will use the lesson strategy; however, other strategies may be used. Accept all reasonable work leading to the correct answer.

Answer A beam bridge has a span of 200 feet.

Discuss What other strategies can you use in addition to working backward to help you solve this problem?

Sample Answer: I can draw a diagram to help me visualize the problem. I can also write an equation to keep track of the operations I need to undo when working backward.

Create Look back at Problem 1. Write and solve a new problem using your own number for the counterweight. Use the strategy *Work Backward* to find the maximum capacity of the elevator car. See teacher notes.

Lesson 1 Strategy Focus: Work Backward 29

In this lesson, students modify Problem 1. They should choose a new weight for the counterweight. If students are struggling, suggest they choose a multiple of 100 to simplify the calculations.

Accept student responses that show a change in the weight of the counterweight and that show how to find the correct solution by working backward.

Independent Practice
Practice

Students should be encouraged to choose any strategy to solve Problems 6 and 7, though many may prefer to use *Work Backward*.

6 Some students may choose to write an equation to solve the problem.

Sample Work

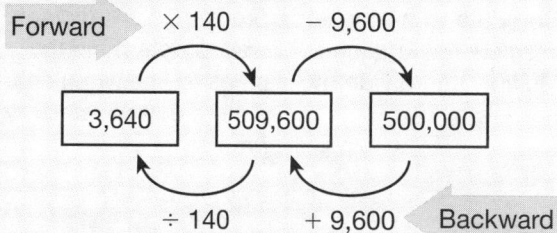

Identify Responses should identify a similar problem and explain, in mathematical terms, how the two problems are alike.

7 If students have trouble getting started, ask them to describe how they would find the total length of the beam bridge if they knew the length of a span of the beam bridge. Then have them reverse the process.

Sample Work

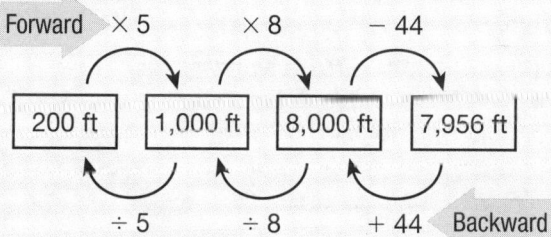

Discuss Students' responses should show that they understand how they can use more than one problem-solving strategy.

Lesson 1   17

Lesson 2

Strategy Focus
Solve a Simpler Problem

Lesson Overview

Lesson Materials: calculator

Skills to Know	Outcome	Math Vocabulary	eResources www.optionspublishing.com
• Write numbers in scientific notation • Compute with numbers in scientific notation • Multiply and divide by powers of 10	Students will recognize that solving a simpler problem is an efficient way to determine the steps to use when solving a more complicated problem.	base, exponent, scientific notation, word equation	• Interactive Whiteboard Transparency 2 • Homework, Unit 1 Lesson 2 • Problem-Solving Checklist, also available in the student worktext, page 7

Modeled Instruction

Learn

To be sure students understand the context of the problem, ask questions like the ones below.

- *Why do you think the problem uses the phrase* closest to the sun *to describe the two distances?*
- *What does it mean to round the first factor to the nearest tenth?*

As students read the problem again, guide them to identify the words and numbers needed to solve the problem.

- *What is 1.471×10^8 in standard form?*
- *Which is the first factor in a number written in scientific notation?*

Reread You may wish to use this Think Aloud to demonstrate how to read a problem to decide which strategy to use.

This problem is about the distance between Earth and the sun and the distance between Mercury and the sun. It tells how far it is from Earth to the sun in scientific notation, 1.471×10^8. It also says that Earth is 3.2 times as far from the sun as Mercury is.

These numbers seem kind of complicated. I wish the problem had just given me whole numbers. Then maybe I could figure out what to do with them. That gives me an idea! I could write a similar problem with easier numbers and figure out how to solve it. Then I could use the same steps with the actual numbers to solve this one.

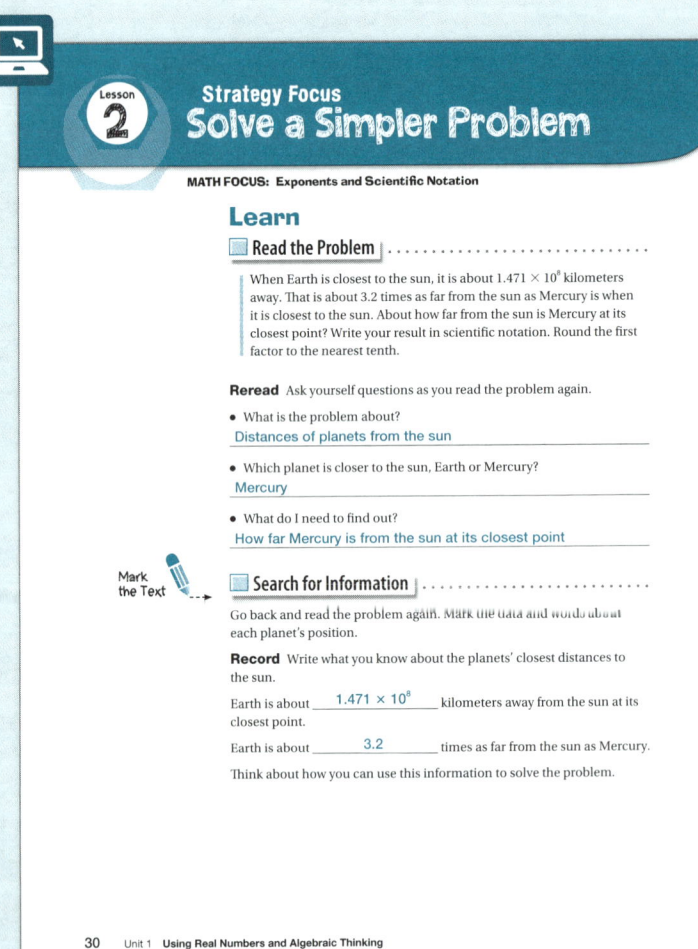

18   Unit 1

Decide What to Do

You know the distance from Earth to the sun at its closest point. You also know that when Earth is at its closest point, it is 3.2 times as far away from the sun as Mercury is from the sun at its closest point.

Ask How can I find how far Mercury is from the sun at its closest point?

- I can use the strategy *Solve a Simpler Problem*.
- I can write an equation using simpler numbers, then substitute the more complicated numbers from the problem to solve it.

Use Your Ideas

Step 1 Write a **word equation** to represent the distances in problem. Then use simpler numbers.

Mercury to the sun = $\frac{\text{Earth to the sun}}{3.2}$

$\frac{6 \times 10^3}{3 \times 10^0} = \frac{6}{3} \times 10^{3-0}$

$= 2 \times 10^{\boxed{3}}$

Think of 3.2 as 3.2×10^0.

Step 2 Use the same method as in Step 1 to divide the actual numbers.

$\frac{1.471 \times 10^8}{3.2} = \frac{1.471}{3.2} \times 10^{\boxed{8}-\boxed{0}}$

$\approx \underline{0.46} \times 10^{\boxed{8}}$

Step 3 Rewrite the number in **scientific notation**. The base of the second factor in scientific notation is always 10. The **exponent** can be positive, negative, or 0.

$0.46 \times 10^8 = 4.6 \times 10^{\boxed{7}}$

Mercury is about ___4.6×10^7___ kilometers away at its closest point to the sun.

Review Your Work

Make sure that you changed the exponent as well as the placement of the decimal point when writing the answer in scientific notation.

Define How did solving a simpler problem help you remember how to compute with numbers written in scientific notation?

Sample Answer: The simpler problem helped me remember to subtract exponents when I divide.

31

Try

Solve the problem.

1. Weights on Jupiter are about 2.14×10^2 times what they are on Earth. If a space module weighs 8.203×10^2 pounds on Earth, what will it weigh on Jupiter? Write your answer in scientific notation. Round the first factor to the nearest hundredth.

Read the Problem and Search for Information

Think about a **strategy** you could use to find a solution.

Decide What to Do and Use Your Ideas

You need to multiply. Use the strategy *Solve a Simpler Problem* to help you find the product of numbers written in scientific notation.

Step 1 Write a simpler problem to understand how to multiply the actual numbers. Round the first factor of the actual numbers to the nearest whole number.

$(2 \times 10^2) \times (8 \times 10^2) = 2 \times 8 \times 10^{2+2}$

$= 16 \times 10^{\boxed{4}}$

$= 1.6 \times 10^{\boxed{5}}$

Step 2 Use the same method as in Step 1 to multiply the actual numbers. Write your answer in scientific notation. Round the first factor to the nearest hundredth.

$(2.14 \times 10^2) \times (8.203 \times 10^2) = 2.14 \times 8.203 \times 10^{2+2}$

$\approx \underline{17.6} \times 10^{\boxed{4}}$

$= \underline{1.76} \times 10^{\boxed{5}}$

So the module will weigh about ___1.76×10^5___ pounds on Jupiter.

Review Your Work

Check that you used the correct exponent in your answer.

Conclude How did solving the simpler problem with rounded numbers in Step 1 help you check your answer?

Sample Answer: The rounded numbers in Step 1 gave an estimate of the actual answer so I could check that my answer was close.

32 Unit 1 Using Real Numbers and Algebraic Thinking

Modeled Instruction (continued)

Help students connect the facts they know with a strategy they can use to solve the problem.

- *Why might using simpler numbers make it easier to understand how to solve the problem?*

Ask questions that encourage students to think critically about the steps in the solution process.

- *Why might you use 6×10^3 and 3×10^0 as the simpler numbers?*
- *Why is 0.46×10^8 not considered to be written in scientific notation?*

Emphasize to students the importance of checking that their answers are written in the correct form.

HOTS Define Students' responses should note that simpler numbers helped them see that they should divide or helped them remember how to divide by powers of 10.

Guided Practice

Try

1. Be sure students understand how solving a simpler problem can help them understand how to solve the original problem.

   Check to see that students understand the information given in the problem.

   - *Where are weights heavier, on Earth or on Jupiter? How do you know?*

   Prompt students to consider the reason for each step.

   - *How do you know that you should multiply?*

   Have students check to see if their answer is correctly written in scientific notation.

   HOTS Conclude Explanations should show an understanding that estimates are a good way to check the reasonableness of an answer.

Lesson 2 19

Scaffolded Practice
Apply

② Ask students how the wording of the information helps them solve the problem.
- *How does the statement* Earth is larger than Mars *help you set up the problem?*
- *Why do you add even though the problem says* 2.98×10^3 *kilometers less than?*

Summarize Explanations should show that students understand that only the decimal factors are added. The power of 10 remains the same.

③ Lead students through the steps of solving a simpler problem.
- *How can a simpler problem help you figure out the steps needed to solve the original problem?*
- *Why is it helpful for the numbers to have the same power of 10 when adding or subtracting?*

Investigate Responses should show that students recognize how to write numbers given in scientific notation in standard form so that they can add or subtract.

④ Prompt students to use the solution to a simpler problem to find the answer to the more complex problem.
- *How could you write a simpler problem that would help you see how to find the answer?*

Examine Students' responses should indicate that they can identify common errors made when working with numbers written in scientific notation.

⑤ Have students check the answer in the context of the original problem.
- *How can you use multiplication to check your answer?*
- *Why are the answers for the suggested simpler problem and the actual problem the same?*

Demonstrate Responses should show an understanding of how to interpret 0 as an exponent.

Apply

Solve the problems.

② Earth is larger than Mars. The radius of Mars at its equator is about 3.4×10^3 kilometers. That is about 2.98×10^3 kilometers less than the radius of Earth at its equator. What is the approximate radius of Earth? Give your answer in scientific notation.

Ask Yourself: Do I need to add or subtract?

Radius of Mars + Amount less than radius of Earth = Radius of Earth
Use simpler numbers.
$$3 \times 10^3 + 2 \times 10^3 = \underline{5 \times 10^3}$$

Hint First solve the problem with simpler numbers like 3×10^3 to help you understand how to find a solution.

Now use the actual numbers.
$$3.4 \times 10^3 + \underline{2.98 \times 10^3} = \underline{6.38 \times 10^3}$$

Answer The radius of Earth is about 6.38×10^3 kilometers.

Summarize What did the simpler problem tell you about how to add numbers in scientific notation when the exponents are the same?

Sample Answer: I can just add the decimal numbers and keep the same power of 10.

③ The moon is 3.844×10^5 kilometers from Earth. Earth is about 1.496×10^8 kilometers from the sun. How much farther is Earth from the sun than from the moon? Write your answer in scientific notation. Round the first factor to the nearest thousandth.

Ask Yourself: How can I rewrite a number given in scientific notation so that it has the same power of 10 as another number?

Earth to sun − Earth to moon = Distance farther
Use simpler numbers. Subtract 3×10^5 from 1×10^8.
Rewrite 3×10^5 as a multiple of 10^8. $3 \times 10^5 = 0.003 \times 10^8$
$1 \times 10^8 - 0.003 \times 10^8 = 0.997 \times 10^8$

Hint 1.0×10^5
$= 0.1 \times 10^6$
$= 0.01 \times 10^7$
$= 0.001 \times 10^8$

Now use the actual numbers.
$$1.496 \times 10^8 - 0.003844 \times 10^8 \approx 1.492 \times 10^8$$

Answer Earth is about 1.492×10^8 kilometers farther from the sun than from the moon.

Investigate What is another way you could have solved the problem besides writing the numbers with the same power of 10?

Sample Answer: I could have written the numbers in standard form, subtracted, then written the difference in scientific notation.

Lesson 2 Strategy Focus: Solve a Simpler Problem 33

---

④ The closest Earth gets to the sun is about 9.14×10^7 miles. At its farthest point, Earth is about 9.45×10^7 miles from the sun. How much farther is Earth from the sun at its farthest point than at its closest point? Write your answer in both scientific notation and in standard form.

Ask Yourself: How do I subtract two numbers in scientific notation when the exponents are the same?

$$9.45 \times \underline{10^7} - 9.14 \times 10^7 = \underline{0.31 \times 10^7}$$

Hint Remember, the first factor has to be greater than or equal to 1 and less than 10 when a number is written in scientific notation.

Answer Earth is about 3.1×10^6, or 3,100,000 miles, farther from the sun at its farthest point than at its closest point.

Examine A student says the answer is 3.1 miles. What mistake might the student have made?

Sample Answer: The student might have left off the powers of 10.

⑤ The mass of Mars is about 6.4×10^{23} kilograms. The mass of Mercury is about 3.3×10^{23} kilograms. About how many times as great is the mass of Mars as the mass of Mercury? Round to the nearest whole number.

Ask Yourself: Do I need to multiply or divide?

$$6.4 \div 10^{23} \div 3.3 \div 10^{23} = \underline{1.9}$$

Hint Try using simpler numbers like 6×10^2 and 3×10^2.

Answer The mass of Mars is about 2 times the mass of Mercury.

Demonstrate What does your simpler problem tell you about the power of 10 in the quotient when you divide two numbers written in scientific notation that have the same power of 10?

Sample Answer: The exponent of 10 in the quotient will be 0. Since 10^0 is 1, the second factor in the quotient is just 1.

34 Unit 1 Using Real Numbers and Algebraic Thinking

20 Unit 1

Practice

Solve the problems. Show your work.

6. An Astronomical Unit (AU) is about 9.3×10^7 miles, or about 1.5×10^8 kilometers. The average distance from Earth to Jupiter is about 4.455 AU. How far is that in miles? Write your answer in scientific notation. Round the first factor to the nearest thousandth.

   Some students will use the lesson strategy; however, other strategies may be used. Accept all reasonable work leading to the correct answer.
   Answer The average distance from Earth to Jupiter is about 4.143×10^8 miles.

   Identify What information is given that is not needed to solve the problem?
   Sample Answer: An Astronomical Unit (AU) is about 1.5×10^8 kilometers.

7. To plan the orbit of a satellite around a planet, a space scientist needs to know the circumference of the planet at its equator. On Earth, that distance is about 2.4901×10^4 miles. The circumference at Jupiter's equator is about 2.79118×10^5 miles. About how much greater is the distance around Jupiter than the distance around Earth? Write your answer in scientific notation. Round the first factor to the nearest hundredth.

   Some students will use the lesson strategy; however, other strategies may be used. Accept all reasonable work leading to the correct answer.
   Answer The distance around Jupiter is about 2.54×10^5 miles greater than the distance around Earth.

   Generalize How can solving a simpler problem help you to compute with numbers in scientific notation?
   Sample Answer: A simpler problem can help me see how to find a solution. It can also help me estimate my answer by figuring out the correct power of 10.

    **Create** Look back at Problem 6. Write and solve a problem using the information about the equivalent distance of an AU in kilometers. Use the strategy *Solve a Simpler Problem* to answer the question. See teacher notes.

   Lesson 2 Strategy Focus: Solve a Simpler Problem   35

 Create In this lesson, students modify Problem 6 by choosing a different distance that must be converted from Astronomical Units. If students are struggling, suggest that they first try to convert a simple distance, such as 2 AU, to kilometers before writing a problem that uses more difficult numbers.

Accept student responses that give a distance in astronomical units and ask for an equivalent distance in kilometers. Students should use the strategy *Solve a Simpler Problem* to help them arrive at a correct solution.

Independent Practice
Practice

Students should be encouraged to choose any strategy to solve Problems 6 and 7, though many may prefer to use *Solve a Simpler Problem.*

6. Some students may choose to write an equation to solve the problem, without first trying a simpler problem.

Sample Work
$$9 \times 10^7 \times 4 = 36 \times 10^7$$
$$= 3.6 \times 10^8$$

$$9.3 \times 10^7 \times 4.455 = 41.4315 \times 10^7$$
$$\approx 4.143 \times 10^8$$

Identify Responses should show that students read the problem carefully and understand that since the question asks for miles, the information about kilometers is not needed.

7. Make sure students understand that the question is asking how *much* greater, not how *many times* greater.

Sample Work
$$3 \times 10^5 - 2 \times 10^4 = 30 \times 10^4 - 2 \times 10^4$$
$$= 28 \times 10^4$$
$$= 2.8 \times 10^5$$

$$2.79118 \times 10^5 - 2.4901 \times 10^4$$
$$= 27.9118 \times 10^4 - 2.4901 \times 10^4$$
$$= 25.4217 \times 10^4$$
$$\approx 2.54 \times 10^5$$

Generalize Students' explanations should note that a simpler problem can help them figure out steps to follow or provide an estimate that helps them to check an answer when working with numbers in scientific notation.

Lesson 2   21

Lesson 3 — Strategy Focus: Write an Equation

Lesson Overview

Lesson Materials: calculator

Skills to Know	Outcome	Math Vocabulary	eResources www.optionspublishing.com
• Solve linear equations in one variable • Solve linear equations in two variables	Students will recognize that equations are an efficient way to represent real-world situations mathematically.	variable	• Interactive Whiteboard Transparency 3 • Homework, Unit 1 Lesson 3 • Know-Find Table • Problem-Solving Checklist, also available in the student worktext, page 7

Modeled Instruction

Learn

To be sure students understand the context of the problem, ask questions such as the ones below.

- *What does it mean to travel upstream and to travel downstream?*
- *How does the current affect the speed of the canoe?*

As students read the problem again, pose questions to help them identify important information.

- *How do the current speed and paddling speed affect the overall speed of each family?*
- *Which speed is the answer to the question?*

 Use a Graphic Organizer You may wish to use this Think Aloud to demonstrate how to analyze the information before starting to solve the problem.

This problem has a lot of information. A Know-Find Table will help me keep track of the information and help me decide how to use it. In the Know column, I can put The paddling speed and speed of the current are the same for each family. *I can also put* The rates that each family traveled *in the Know column. In the Find column, I can put* The speed of the current.

I think I can write and solve equations to help me answer the question.

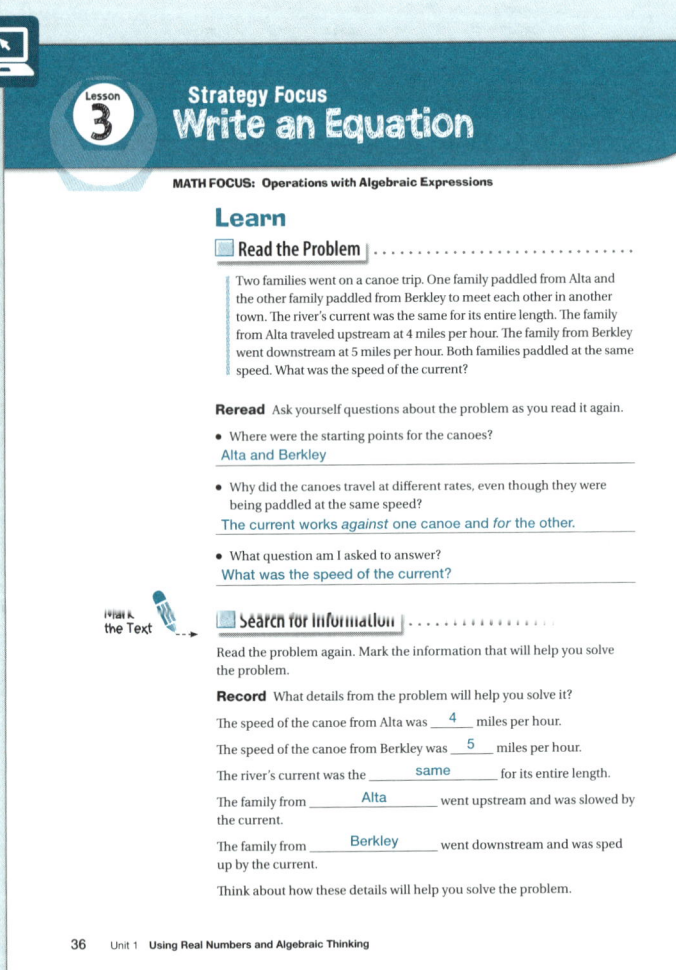

22   Unit 1

Decide What to Do

You know the rate and direction each canoe traveled. You know that the current was the same for the river's entire length.

Ask How can I find the speed of the current?

- I can use the strategy *Write an Equation* to write equations for each situation.
- Next, I can solve one of the equations for one of the variables and then substitute the resulting expression into the other equation.

Use Your Ideas

Step 1 Write equations for the two speeds.

Let p be the speed that the families paddled, in miles per hour.
Let c be the speed of the current, in miles per hour.

$p - c = \underline{4}$
$p + c = \underline{5}$

> You can use equations to describe the two different speeds.

Step 2 Solve the first equation for p.

$p - c = 4$
$p - c + c = 4 + c$ ← Add c to both sides.
$p = 4 + c$ ← Simplify.

Step 3 Substitute $4 + c$ for p in the second equation.

$p + c = 5$
$(4 + c) + c = 5$ ← Substitute.
$4 + 2c = 5$ ← Combine like terms.
$4 - 4 + 2c = 5 - 4$ ← Subtract 4 from both sides.
$2c = 1$ ← Simplify.
$c = \frac{1}{2}$ ← Divide both sides by 2.

So the speed of the current is $\underline{\frac{1}{2}}$ mile per hour.

Review Your Work

Check to see that you correctly labeled your answer.

Define How can you use your answer to the problem to find p, the speed that the families paddled?

Sample Answer: I know that $p - c = 4$ and that $c = \frac{1}{2}$.
So $p = 4 + \frac{1}{2}$. The families paddled at $4\frac{1}{2}$ miles per hour.

Try

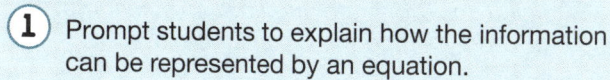

Solve the problem.

① When Patty climbs cliffs, she needs a length of rope equal to twice the height of the cliff she plans to climb. She also likes to have a safety factor of 15 extra feet on each end of the rope. If she brings a rope that is 240 feet long, how high is the cliff Patty plans to climb?

Read the Problem and Search for Information

Think about how the rope length relates to the height of the cliff.

Decide What to Do and Use Your Ideas

You can use the strategy *Write an Equation*.

Step 1 Write a word equation.

2 × (Height of cliff + Safety factor) = Length of rope

Step 2 Write a symbolic equation and solve the problem.
Let h be the height of the cliff.

$2(h + \underline{15}) = \underline{240}$ ← Use the Distributive Property.
$2h + \underline{30} = \underline{240}$
$2h + \underline{30 - 30} = \underline{240 - 30}$ ← Subtract 30 from both sides.
$2h = \underline{210}$ ← Simplify.
$2h \div 2 = \underline{210 \div 2}$ ← Divide both sides by 2.
$h = \underline{105}$ ← Simplify.

So the cliff is $\underline{105\text{ feet}}$ high.

> **Ask Yourself**
> There are 15 extra feet of rope on each end, so both the height of the cliff and the safety factor must be doubled.

Review Your Work

Substitute your answer in the words of the original problem to check your work.

Clarify How did you use the fact that the length of the rope needs to be twice the height of the cliff when writing your equation?

Sample Answer: I know *twice the height* means I need to multiply by 2. I used parentheses to show that I wanted to multiply $h + 15$ by 2, and not just multiply h or 15 by 2.

Scaffolded Practice
Apply

2 Prompt students to use variables to represent the situation.
- How many variables do you need?
- How would you represent the amount the company received for half-day trips? For full-day trips?

HOTS Sequence Students' explanations should show that they understand the relationship between the number of half-day trips, full-day trips, and the total number of trips.

3 Guide students to check that their answers are reasonable in the context of the problem.
- Will your answer be more or less than $48? Why?
- Once you have found the answer, how can you use the information from the problem to make sure it is correct?

HOTS Interpret Explanations should show an understanding that $\frac{1}{3}$ off the total is $\frac{2}{3}$ of the price.

4 Remind students to use the distance formula, $d = r \times t$.
- If you know the rate and time an object moves, how do you find the distance it moves?

HOTS Explain Responses should note that miles per hour means $\frac{miles}{hour}$ and that $\frac{miles}{hour} \times hours = miles$.

5 Ask questions that focus students on the information given in the problem.
- If the mean high temperature for 7 days was 0°C, what was the sum of the temperatures on those 7 days?

HOTS Analyze Students' responses should show an understanding that the scientist recorded 7 values to calculate the mean for one week, and Rita would only be correct if only 2 values were recorded.

Apply
Solve the problems.

2 At a rafting company, a half-day trip costs $50. A full-day trip costs $100. There were 16 trips today. How many of each type of trip were there if the income for the day was $1,300?

Let h be the number of half-day trips. Let f be the number of full-day trips.

$h + f = 16$ $\underline{50}\ h + \underline{100}\ f = 1,300$

Hint: To answer the question this problem is asking, you need to find the value of both variables.

From the first equation, you know that $h = 16 - f$.

$50(16 - f) + \underline{100f} = 1,300$ ← Substitute $16 - f$ for h.

$\underline{50} \times \underline{16} - 50f + 100f = 1,300$ ← Use the Distributive Property.

$\underline{800} + \underline{50f} = 1,300$ ← Simplify.

$50f = \underline{500}$

Ask Yourself: If I know the value of f, how can I find the value of h?

Answer There were 6 half-day and 10 full-day trips.

Sequence What steps did you take after you found the value of f?
Sample Answer: I subtracted 10 from 16 to find the number of half-day trips.

3 Before hiking, Mr. Chan buys 3 carabiners, which are metal clips that hold objects. He also buys socks that normally cost $15. Mr. Chan receives $\frac{1}{3}$ off all purchases from the store because he is a guide. He pays $48. What is the price of each carabiner for people who are not guides?

Hint: Let c be the price of one carabiner for people who are not guides.

$\frac{2}{3}$ (price of 3 carabiners + socks) = $\underline{\text{Mr. Chan's cost}}$

$\frac{2}{3}(3c + 15) = \underline{48}$

$\frac{2}{3} \times \underline{3c} + \frac{2}{3} \times \underline{15} = \underline{48}$

Ask Yourself: What equation will describe the situation?

Answer The price of each carabiner is $19.

Interpret Why does the equation use the number $\frac{2}{3}$ when the number in the problem is $\frac{1}{3}$?
Sample Answer: Mr. Chan gets $\frac{1}{3}$ off the price, which is $\frac{2}{3}$ of the price.

Lesson 3 **Strategy Focus: Write an Equation** 39

---

4 An Autonomous Underwater Vehicle (AUV) spent 8 hours underwater. Part of the time, it moved at 2.5 miles per hour. The rest of the time, it moved at 1 mile per hour. It traveled 17 miles in all. How much time did the AUV spend at each speed?

Ask Yourself: I can write an equation to find the times. How can I represent miles when I know miles per hour and hours?

Time at first speed + Time at second speed = Total time

Time A + Time B = $\underline{8}$ hours

Speed A × Time A + Speed B × Time B = Total miles

$\underline{2.5} \times$ Time A + $\underline{1} \times$ Time B = $\underline{17}$ miles

Hint: Write two equations. Solve one equation. Then substitute the resulting expression into the other equation.

Answer The AUV spent 6 hours at 2.5 miles per hour and 2 hours at 1 mile per hour.

Explain Why do you get *miles* when you multiply *miles per hour* by *hours*?
Sample Answer: Miles per hour is *miles* divided by *hours*. If I first divide, then multiply miles by hours, I will get miles.

5 A scientist at McMurdo Station in Antarctica finds and records the mean high temperature for the week. Today is the last day of the week. For the first six days of the week, the mean high temperature was −1° C. After finding today's high temperature, she found that the mean was 0° C for the week. What was today's high temperature?

Hint: Mean is $\frac{\text{sum of numbers}}{\text{number of numbers}}$.

Ask Yourself: How can I find the sum of the high temperatures for the first six days of this week?

$\frac{\text{Number of days} \times \text{Mean temperature} + \text{New temperature}}{\text{New number of days}}$ = New mean

$\frac{6 \times \underline{-1} + \underline{t}}{7} = \underline{0}$

Answer Today's high temperature was 6° C.

Analyze Rita thinks today's high temperature was 1° C. How do you know that answer cannot be correct?
Sample Answer: It would be correct only if the average was for 2 days instead of 7.

Practice

Solve the problems. Show your work.

6 C-17 cargo planes carry freight and passengers in and out of McMurdo Station, Antarctica. Suppose that 57 flights flew to McMurdo Station in one year. Each flight had a crew of six people. Altogether, 2,622 people traveled on these flights. What was the average number of non-crew people on each flight?

Some students will use the lesson strategy; however, other strategies may be used. Accept all reasonable work leading to the correct answer.

Answer An average of 40 non-crew people were on each flight.

Defend Describe how you can solve this problem using equations and 2 variables.

Sample Answer: I can let t be the total number of non-crew people and n be the average number. So $t = 2,622 - (57 \times 6)$. Because $n = \frac{t}{57}$, I can substitute $2,622 - (57 \times 6)$ into the second equation and solve for n.

7 Mr. Abrams and Mr. Browning are following Robert Scott's 1912 route to the South Pole. They are pulling sleds with all 600 pounds of their gear. The gear on Mr. Abrams's sled is worth $75 per pound. The gear on Mr. Browning's sled is worth $100 per pound. The total value of all of the gear is $52,500. What is the weight of the gear on each sled?

Some students will use the lesson strategy; however, other strategies may be used. Accept all reasonable work leading to the correct answer.

Answer 300 pounds of gear are on each sled.

Generalize How do you decide what to do when you have the same two variables in two equations?

Sample Answer: I look for a way to write an expression for one variable. Then I substitute that expression into the equation for the other variable.

 Look back at Problem 2. Change the total number of trips and the total income for the day. Write and solve your new problem.

Create See teacher notes.

Lesson 3 **Strategy Focus: Write an Equation** 41

 **Create**

In this lesson, students modify Problem 2 by changing the total number of trips and the total income. If students are struggling, suggest they decide upon numbers for the answers first and then determine the new numbers to use in the problem.

Accept student responses that provide a problem that can be solved using an equation or equations and that include a correct solution.

Independent Practice
Practice

Students should be encouraged to choose any strategy to solve Problems 6 and 7, though many may prefer to use *Write an Equation*.

6 Some students may choose to use *Guess, Check, and Revise* to solve the problem.

Sample Work

Let x = the number of non-crew people on each flight.

$57(x + 6) = 2,622$

$57x + 342 = 2,622$

$57x = 2,280$

$x = 40$

 Defend Students' explanations should show the steps for solving this problem using two variables.

7 Once students have solved the equation, have them read the problem again to make sure they answered the question completely.

Sample Work

Let a = the weight on Mr. Abrams' sled and b = the weight on Mr. Browning's sled.

$a + b = 600$, so $a = 600 - b$

$75a + 100b = 52,500$

$75(600 - b) + 100b = 52,500$

$45,000 - 75b + 100b = 52,500$

$25b = 7,500$

$b = 300$

$a + 300 = 600$

$a = 300$

 Generalize Responses should mention substituting an expression for one variable, or eliminating one variable when combining the two equations.

Lesson 3   25

Lesson 4: Strategy Focus — Use Logical Reasoning

Lesson Overview

Lesson Materials: calculator

Skills to Know	Outcome	Math Vocabulary	eResources www.optionspublishing.com
• Solve equations • Solve inequalities	Students will recognize that logical reasoning is an efficient way to find relationships among pieces of information given in a problem.	inequality	• Interactive Whiteboard Transparency 4 • Homework, Unit 1 Lesson 4 • Problem-Solving Checklist, also available in the student worktext, page 7

Modeled Instruction

Learn

To be sure students understand the context of the problem, ask questions such as the ones below.

- What does the phrase *evenly distributed* mean?
- What do you know about the rainfall in the months other than January and May?

As students read the problem again, pose questions to help them identify important information.

- Which months have about the same amount of rainfall?

 Represent You may wish to use this Think Aloud to demonstrate how to organize the information so that it is easier to use.

The problem is about rain in a tropical rainforest. I know the expected amount of rain in May and information about the rain in other months. I think I will make a table to help me organize what I know. I can write that the rainfall in May is more than 13 inches. The rainfall for January should be 4 inches more than the rainfall in May. The other months should all have the same amount of rainfall as May.

Now that I can see how much it rains each month, I can solve the problem.

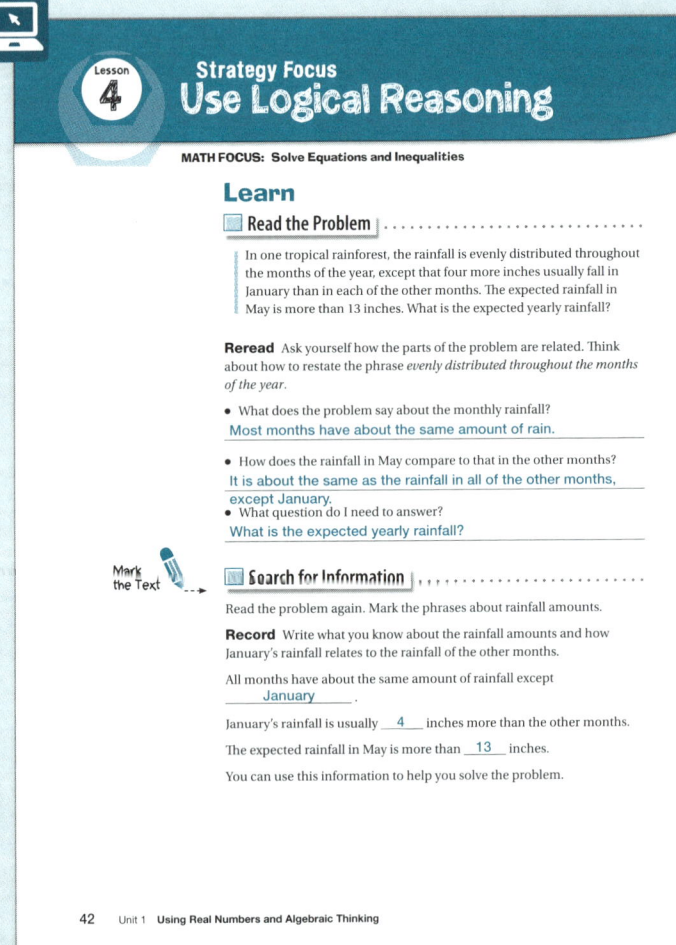

26 Unit 1

Decide What to Do

You know the relationship between the usual rainfall in January and in the other months. You also know the expected rainfall in May.

Ask How can I find the total expected rainfall for one year?

- I cannot simply multiply the rainfall expected in May by the number of months in a year because January has more rain.
- I can use the strategy *Use Logical Reasoning* to help write an **inequality** to describe the problem situation. Then I can solve my inequality and check that it describes the result.

Use Your Ideas

Step 1 Write an inequality to describe the situation. Fill in what you know. Use y to represent the yearly rainfall, in inches.

$$\frac{\text{yearly rain} - \text{extra rain in January}}{\text{number of months in a year}} > \text{expected May rainfall}$$

$$\frac{(y - \underline{4})}{12} > 13$$

The words *more than* tell you that you need to use > in your inequality.

Step 2 Solve the inequality.

$y - \underline{4} > 12 \times \underline{13}$

$y - \underline{4} > \underline{156}$

$y > \underline{160}$

So the expected yearly rainfall is more than $\underline{160}$ inches.

Review Your Work

Check by testing values above and below 160 in the inequality.

Recognize How did logical reasoning help you write the inequality?

Sample Answer: Logical reasoning helped me identify the relationships between the numbers. It also helped me decide which inequality symbol to use.

43

Modeled Instruction (continued)

Help students see how the details they have identified can be used to choose a problem-solving strategy.

- *How can you represent all the given information with a mathematical sentence?*

Ask questions that encourage students to think critically about the steps in the solution process.

- *Why do you subtract the extra rain in January when the word* extra *means more?*
- *Why do you divide by 12 when only 11 months have the same rainfall?*

Emphasize to students that you can check an inequality by testing values on each side of the inequality.

- *What would the rainfall be in May if there were 172 inches of rain in a year? Would that work in the problem?*

HOTS Recognize Explanations should show how to use relationships between the numbers in the problem to write the inequality.

Try

Solve the problem.

① Some scientists say the rainforest is disappearing at a rate of 1.5 acres every second. If they are correct, how many acres of rainforest are disappearing each day?

Mark the Text

Read the Problem and Search for Information

Restate the problem in your own words. Think about what you know that you can use to solve the problem.

Decide What to Do and Use Your Ideas

You can use logical reasoning. Think about the units of time.

Ask Yourself
What do I know about how seconds are related to days?

Step 1 List the relationships between units of time.

1 day = $\underline{24}$ hours

1 hour = $\underline{60}$ minutes

1 minute = $\underline{60}$ seconds

Step 2 Work through the problem logically, using the relationships between units of time.

$$\frac{1.5 \text{ acres}}{\text{second}} \times \frac{60 \text{ seconds}}{\text{minute}} = \frac{90 \text{ acres}}{\text{minute}}$$

$$\frac{90 \text{ acres}}{\text{minute}} \times \frac{60 \text{ minutes}}{\text{hour}} = \frac{5,400 \text{ acres}}{\text{hour}}$$

$$\frac{5,400 \text{ acres}}{\text{hour}} \times \frac{24 \text{ hours}}{\text{day}} = \frac{129,600 \text{ acres}}{\text{day}}$$

So $\underline{129,600}$ acres of rainforest are disappearing every day.

Review Your Work

Make sure that your units were set up properly when multiplying.

Demonstrate How would you use this method to find the number of acres that disappear in the month of April? Explain.

Sample Answer: I would multiply my answer by 30 because there are 30 days in April.

44 Unit 1 Using Real Numbers and Algebraic Thinking

Guided Practice

Try

① Encourage students to think about the problem logically.

Help students identify the relationship between the units of time needed to solve the problem.

- *How will you use what you know about units of time to solve the problem?*

Prompt students to explain each ratio in Step 2.

- *Why is seconds in the denominator of the first ratio and in the numerator of the second?*
- *Why can you multiply by $\frac{60 \text{ seconds}}{\text{minute}}$ without changing the rate?*

Suggest students check the ratios that they used to change from one unit of time to another.

HOTS Demonstrate Students should show they understand how to use the fact that there are 30 days in April.

Lesson 4 27

Scaffolded Practice
Apply

2 Guide students through the steps needed to represent the situation with an equation.

- If t represents Earth's total land area, what expression represents the area covered by rainforest?

HOTS Explain Students' responses should show an understanding that the answer must be many times greater than 8,898,000 square kilometers.

3 Ask questions to help students understand how the information in the problem is related.

- Is the amount spent on smallpox more than or less than the amount spent on polio?
- What does $\frac{8}{100}$ refer to?

HOTS Determine Explanations should show that students can determine which information is needed to answer the question. They should note that the number of polio cases is not needed to solve the problem.

4 Familiarize students with the idea of three-part inequalities.

- What does $a \leq b \leq c$ mean? What are some possible values of a, b, and c?

HOTS Interpret Responses should show that students see how the information can be used in other ways.

5 Have students reflect on how this problem can be solved in a way similar to the way Problem 1 was solved.

- How could you break this problem into two steps?

HOTS Discuss Students' responses should demonstrate an understanding that they should multiply since the cubic meter is the larger unit. They may relate this to more basic units, such as multiplying by 12 to find how many inches are in a number of feet.

28   Unit 1

Apply

Solve the problems.

2 About 8,898,000 square kilometers of Earth are covered by tropical rainforest. This leaves $\frac{47}{50}$ of Earth's land area not covered by tropical rainforest. What is Earth's total land area?

Ask Yourself: What fraction of Earth is covered by rainforest?

If $\frac{47}{50}$ of Earth is not covered by rainforest, then $\frac{3}{50}$ is covered by rainforest.

Area of Earth × Rainforest fraction = Area covered by rainforest

$$t \times \frac{3}{50} = 8{,}898{,}000$$
$$t = 148{,}300{,}000$$

Hint: Think of Earth as 1 whole.

Answer Earth's total land area is about 148,300,000 square kilometers.

Explain How do you know that your answer makes sense?

Sample Answer: Since $\frac{3}{50}$ is a small part of the whole, my answer must be much greater than 8,898,000 square kilometers.

3 The program to eliminate smallpox cost about 300 million dollars. That is less than $\frac{8}{100}$ of what has been spent to eliminate polio. There are still 3,500 polio cases each year. What is the least amount that could have been spent trying to eliminate polio so far?

Ask Yourself: Logically, will my answer be an exact amount of money?

Smallpox amount < Polio amount × Fraction of polio amount

300,000,000 < p × 0.08

3,750,000,000 < p

Hint: Write an inequality to model the situation.

Answer The least amount that could have been spent trying to eliminate polio so far is $3,750,000,000.

Determine What information is given that is not needed?

Sample Answer: There are still 3,500 cases of polio each year.

Lesson 4   Strategy Focus: Use Logical Reasoning   45

Hint: You can use *million* as a unit label. That will make it easier to compute with the numbers in the problem than if you wrote them out in standard form.

4 A state park system is planning its budget. It costs between 32 and 34 million dollars each year to run the park system well. The state budgeted 7.5 million dollars to run the park system this year. By what factor would the budget need to be multiplied to ensure the park system runs well? Round to the nearest tenth.

low estimate ≤ 2010 amount × factor ≤ high estimate

$$32 \leq 7.5 \times f \leq 34$$

Ask Yourself: Is there only one correct answer or can I give my answer as a range?

Answer The 2010 budget would need to be multiplied by at least 4.3, but not more than 4.5.

Interpret What is another question you could ask from the information in the problem?

Sample Answer: About how much more money does the park system need to run well?

Hint: There are about 3.6 cords in a cubic meter.

5 Harvesting teak from rainforests causes harm to other plants. One way to prevent this is to grow teak on plantations. A teak plantation can produce about 20.4 cubic meters of wood per month. People often buy wood by the cord. How many cords per year could the teak plantation produce? Round to the nearest cord.

Ask Yourself: How can I use logical reasoning to write cubic meters per month as cords per year?

$$\frac{20.4 \text{ m}^3}{\text{month}} \times \frac{3.6 \text{ cords}}{\text{m}^3} \times \frac{12 \text{ months}}{\text{year}} = \frac{881.28 \text{ cords}}{\text{year}}$$

Answer The teak plantation could produce about 881 cords per year.

Discuss Why do you multiply rather than divide to find the number of cords?

Sample Answer: According to the hint, there are more cords than cubic meters, so to change from cubic meters, I needed to multiply.

46   Unit 1   Using Real Numbers and Algebraic Thinking

Practice

Solve the problems. Show your work.

6 A rectangular section of rainforest is 1,000 meters by 2,500 meters. Large land areas are sometimes expressed in hectares. One square kilometer is 100 hectares. What is the area of the rectangular section in hectares?

Some students will use the lesson strategy; however, other strategies may be used. Accept all reasonable work leading to the correct answer.

Answer The area of the rectangular section is 250 hectares.

Analyze Tyler says that the area of the section is 250 million hectares. What mistake could he have made?

Sample Answer: Tyler might not have first changed meters to kilometers.

7 Twenty years ago, there were more than 350,000 cases of polio each year. Now, because of programs to eliminate polio, there are about $\frac{1}{100}$ as many cases per year. About how many more polio cases each year were there 20 years ago than there are now?

Some students will use the lesson strategy; however, other strategies may be used. Accept all reasonable work leading to the correct answer.

Answer There were about 346,500 more cases each year of polio 20 years ago than there are today.

Examine Which words in the first sentence tell you that you cannot give an exact answer?

Sample Answer: The words *more than* tell me that the answer will not be exact.

Create Look back at Problem 5. Change the rate at which the plantation produces wood. Write and solve a new problem to find the number of cords than can be produced in a different amount of time. See teacher notes.

Lesson 4 Strategy Focus: Use Logical Reasoning 47

In this lesson, students are asked to modify Problem 5 by changing the rate at which the plantation produces wood. If students are struggling, suggest they first give a whole number for the number of cubic meters of wood per year before writing a problem using a decimal amount for a different time period.

Accept student responses that give a new rate for the amount of wood the plantation produces in a given time and that provide a correct solution.

Independent Practice
Practice

Students should be encouraged to choose any strategy to solve Problems 6 and 7, though many may prefer to use *Use Logical Reasoning*.

6 Some students may write an equation to find the area in square meters first, then rewrite that number in square kilometers. Others may first rewrite meters as kilometers before finding the area.

Sample Work

area = $\frac{1{,}000 \text{ m}}{1} \times \frac{1 \text{ km}}{1{,}000 \text{ m}} \times \frac{2{,}500 \text{ m}}{1} \times \frac{1 \text{ km}}{1{,}000 \text{ m}}$

= $2.5 \text{ km}^2 \times \frac{100 \text{ ha}}{1 \text{ km}^2}$

= 250 hectares

Analyze Students' responses should demonstrate that they can identify common errors made when converting measures.

7 Make sure students understand the information given and can identify what the question is asking.

Sample Work

Let d = the difference between the number of cases 20 years ago and now.

$d > 350{,}000 - \left(\frac{1}{100} \times 350{,}000\right)$

$d > 350{,}000 - 3{,}500$

$d > 346{,}500$

Examine Students' explanations should show that they can recognize words such as *about* that tell them they cannot give an exact answer.

Lesson 4 29

UNIT 1 Review

UNIT 1 Review

In this unit, you worked with four problem-solving strategies. You can use many different strategies to solve a single problem. If a strategy does not seem to be working, try a different one.

Problem-Solving Strategies
✓ Work Backward
✓ Solve a Simpler Problem
✓ Write an Equation
✓ Use Logical Reasoning

Check students' work throughout.
Students' choices of strategies may vary.
Solve each problem. Show your work. Record the strategy you use.

1. The mean temperature in August in Antarctica is −28°C. The mean is 25° higher in January than in August. What is the mean January temperature?

   Answer: _The mean January temperature is −3°C._
   Strategy: _Possible Strategy: Write an Equation_

2. The volume of Earth is about 1.0832×10^{12} km³. The volume of Saturn is about 7.636×10^{2} times the volume of Earth. What is the approximate volume of Saturn? Write your answer in scientific notation. Round the first factor to the nearest hundredth.

   Answer: _The volume of Saturn is about 8.27×10^{14} km³._
   Strategy: _Possible Strategy: Solve a Simpler Problem_

3. In Washington State, the maximum time between bridge inspections is 72 months. If a bridge is reported to be in poor condition, the maximum time between inspections is one sixth of that time. All wooden bridges, regardless of condition, are inspected half as often as bridges in poor condition. How often are wooden bridges inspected?

   Answer: _Wooden bridges are inspected every 24 months._
   Strategy: _Possible Strategy: Use Logical Reasoning_

4. A total of 54 students and adult chaperones are going to a concert. Concert tickets cost $8 for students and $10 for adults. The total cost of the tickets is $460. How many adults and how many students are going to the concert?

   Answer: _There are 14 adults and 40 students going to the concert._
   Strategy: _Possible Strategy: Write an Equation_

5. Robert Scott led an expedition team to the South Pole in 1912. During the 78-day trip, the team covered more than 850 miles. What was the team's average daily progress?

   Answer: _The average daily progress was at least 10.9 miles per day._
   Strategy: _Possible Strategy: Use Logical Reasoning_

   Explain how you decided which inequality to use.

   Sample Answer: Since the trip of *at least* 850 miles took 78 days, then the average daily progress must have been at least $\frac{850}{78}$ miles, or $p \geq \frac{850}{78}$.

48 Unit 1 Using Real Numbers and Algebraic Thinking

49

Support for Assessment

The problems on pages 48–51 reflect strategies and mathematics students used in the unit.

Although this unit focuses on four problem-solving strategies, students may use more than one strategy to solve the problems or use strategies different from the focus strategies. Provide additional support for those students who need it.

Work Backward For Problems 7 and 9, ask students to think about the order in which they need to perform the mathematical operations. For Problem 7, review the formula for the volume of a prism if necessary.

Solve a Simpler Problem For Problems 2 and 10, suggest students begin by substituting whole number factors times powers of 10 to figure out the strategy and steps they will use to solve the actual problem.

Write an Equation For Problems 1, 4, and 8, ask students to write an equation in words before substituting numbers and variables. For Problems 4 and 8, ask, *What information will you use to write one equation? What information will you use to write another?*

Use Logical Reasoning For Problem 3, ask students to explain what *half as often* and *one sixth of that time* mean in the context of the problem. For Problem 5, have students explain how the words *at least* affect the answer. For Problem 6, ask students to determine what ratios are needed to convert miles to inches.

You may wish to use the *Review* to assess student progress or as a comprehensive review of the unit.

30

Promoting 21st Century Skills

Write About It
Communication

When students describe what fractions mean in a given context, they are able to reflect on how math is used in real-life examples.

Team Project: Draw the Solar System
Collaboration: 3–4 students

Remind students that in a group project, each person must be a part of both the decision-making process and the work. Monitor the groups to make sure all students are participating. Remind students to use a reasonable scale so that the drawings are small enough to fit on the poster, but large enough for others to see.

Ask questions that help students summarize their thinking. *Do the distances represented on your diagram reflect those in the table? The distance from Saturn to the sun is about 10 times the distance from Earth to the sun. Do the distances on your diagram show this?*

Extend the Learning
Media Literacy

If you have Internet access, navigate to sites where students can find sizes of the planets in order to make three-dimensional models in which the relative sizes are accurate.

🔍 Search **size of planets**

Solve each problem. Show your work. Record the strategy you use.

6. An engineer calculated that building a new road in her local state park will cost $10 per inch. The road will be 2.5 miles long. How much will it cost to build the road?
(Hint: 1 mile = 5,280 feet)

Answer The road will cost $1,584,000.

Strategy *Possible Strategy:* Use Logical Reasoning

7. The roadbed of a bridge over a small creek is a concrete rectangular prism. Its depth is two feet, and its length is twice its width. The volume of the roadbed is 100 cubic feet. How long is the bridge?

Answer The bridge is 10 feet long.

Strategy *Possible Strategy:* Work Backward

8. Anna's hot air balloon company charges $12 more for a ride than Boyd's company. One month, Anna had 25 customers and Boyd had 100. Together, they were paid $8,800 for their rides. How much does each company charge?

Answer Anna's company charges $80 and Boyd's company charges $68.

Strategy *Possible Strategy:* Write an Equation

Explain how you solved the problem.

Sample Answer: I let A be the amount Anna charges and B be the amount Boyd charges, in dollars. I wrote two equations: $A - B = 12$ and $25A + 100B = 8,800$. So $A = 12 + B$. I substituted that expression for A in the second equation and solved for B. Then I substituted the value of B, $68, into the first equation to find A, $80.

9. Region B of a rainforest has five fewer tree species than Region A. Region C has 36 tree species. That is three times as many species as Region B. How many species are in Regions A and B?

Answer Region B has 12 species and Region A has 17 species.

Strategy *Possible Strategy:* Work Backward

10. The surface area of Earth is about 13.4 times the surface area of Earth's moon. The moon's surface area is about 3.79×10^7 km². What is the surface area of Earth? Write your answer in scientific notation. Round the first factor to the nearest tenth.

Answer The surface area of Earth is about 5.1×10^8 km².

Strategy *Possible Strategy:* Solve a Simpler Problem

Write About It

Look back at Problem 3. Two of the periods of time in that problem are expressed as fractional amounts; *one sixth of that time* and *half as often*. Describe what those terms mean in the context of the problem.

Sample Answer: The phrase *one sixth of that time* refers to 72 months, or 6 years. One sixth is 12 months, or 1 year. The phrase *half as often* refers to 1 year. Something that happens half as often as every year happens every 2 years, or 24 months.

Team Project: Draw the Solar System

Your team will make a diagram of the solar system. You want the distances on the diagram to be to scale.

Plan
1. Decide if you will show the planets in a straight line or in different positions around the sun.
2. Think about how large you want your diagram to be. Choose an appropriately sized sheet of paper.

Calculate As a group, choose a scale for your drawing. Use the scale to convert the actual distances of the planets from the sun to the distance they will be in the diagram.

Create Draw your diagram. Be sure you include your scale on the drawing.

Present As a group, show your diagram to the class. Explain how you used the scale to find where to place each planet.

Planet	Distance from the sun (km)
Mercury	5.8×10^7
Venus	1.1×10^8
Earth	1.5×10^8
Mars	2.3×10^8
Jupiter	7.8×10^8
Saturn	1.4×10^9
Uranus	2.9×10^9
Neptune	4.5×10^9

UNIT 2 Problem Solving Using Proportional Reasoning

CCSS 7.RP Ratios and Proportional Relationships

Unit Overview

Lesson	Problem-Solving Strategy	Math Focus
5	Make a Table	Rates, Ratios, and Direct Variation
6	Write an Equation	Percents
7	Work Backward	Percent Applications
8	Write an Equation	Percent Change

Promoting Critical Thinking

Higher order thinking questions occur throughout the unit and are identified by this icon: 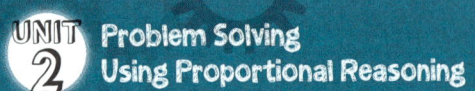. These questions progress through the cognitive processes of remembering, understanding, applying, analyzing, evaluating, and creating to engage students at all levels of critical thinking.

UNIT 2 Problem Solving Using Proportional Reasoning

Unit Theme: Money At Work

Money is a part of everyday life. You use money to pay for things. You can earn money by babysitting or helping out around the house. Many people work together to raise money for various causes. In this unit, you will see how math plays an important role in using and earning money.

Math to Know

In this unit, you will apply these math skills:
- Solve problems involving ratios, rates, proportions, and percents
- Solve problems with percents greater than 100% and less than 1%
- Find percent of increase and percent of decrease in given situations

Problem-Solving Strategies
- Make a Table
- Write an Equation
- Work Backward

Link to the Theme

Write another paragraph about Kareem's conversation with his customer. Include some of the words and numbers from the table.

Kareem receives a phone call about his dog-walking business. He talks to the customer about the different services he offers and their prices.

Students' paragraphs will vary, but should include some words and numbers from the table.

Number of Dogs	30-Minute Walk	60-Minute Walk
1	$12	$20
2	$18	$25

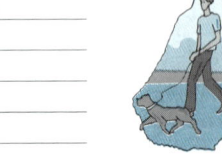

Use Math Language

Review Vocabulary

The list below shows vocabulary terms in this unit. Knowing the meaning of these terms will help you understand the problems.

budget   percent equation   proportion   ratio
interest   profit   rate   tax

Vocabulary Activity   Words in Context

Context can help you determine the meaning of a word. Use words from the list above to complete each sentence.

1. They did not include the purchase of a new car in their household __budget__ this year.
2. She had to pay back the borrowed money plus __interest__.
3. He was lucky to be able to sell the house for a __profit__.
4. When you eat at a restaurant, a sales __tax__ is usually added to the price of food and beverages.

Graphic Organizer   Word Web

Complete the graphic organizer.
- In the center oval, write a definition of *ratio*.
- In each linked oval, write a different vocabulary term from above related to *ratio*. Then write a definition of each term.

Sample Answers:

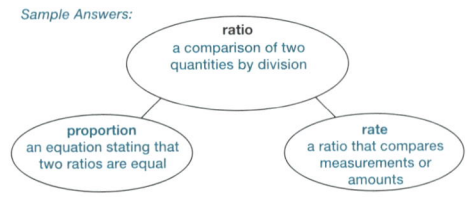

52

53

Link to the Theme   Money at Work

Ask students to read the direction line and story starter. If students are having trouble getting started, ask questions such as, *Why do you think the cost of walking 2 dogs is not double the cost of walking 1 dog? What is the total cost of a daily 30-minute walk for one week?*

Unit 2   Differentiated Instruction

Extra Support

Some students may benefit from activities that strengthen their understanding of percent.

Percents to Decimals Remind students that *percent* means "per 100." One way to change a percent to a decimal is to divide by 100. Point out that a shortcut for dividing by 100 is to move the decimal point two places to the left. Demonstrate the process with a variety of percents such as 52%, 2%, 0.5%, 215%, and 15.5%. Emphasize that the result is not always a decimal between 0.01 and 0.99.

Parts, Wholes, and Percents Demonstrate how to translate word problems into percent equations. Begin by writing the percent equation as *whole* $\times$ % = *part*. Remind students that when translating word problems to equation form, the word *is* usually means "equals," *of* usually means "times," and a variable can stand for the word *what*. Then present these examples:

What is 60% of 120? $\Rightarrow n = 0.6 \times 120$

20% of what number is 100? $\Rightarrow 0.2 \times n = 100$

16 is what percent of 80? $\Rightarrow \quad 16 = \frac{n}{100} \times 80$

Ask students to create their own percent problems using the percent equation.

Challenge Early Finishers

Challenge early finishers to work with problems that are missing the question, such as the one below.

> Usually, 1,200 students attend the high school. During a flu epidemic, the attendance at the school decreased by 30%. Two weeks later, it increased by 30%.

Ask students to determine what they can figure out from the problem. Then have them write a similar question and solve their new problem.

English Language Learners

Vocabulary

Terms About Money Familiarize students with some of the financial terms mentioned throughout the unit. Write the following terms from the unit on the board: *invest, earn, save, account, allowance, deposit, expenses,* and *deducted*. Discuss these terms with students. Then explain that knowing the meaning of these terms can help students understand other forms of these words in the unit, such as *investment, earnings, savings,* and *deductions*. Discuss the meaning of these terms and how they relate to the first set of words. For example, *investment* is a noun form of the verb *invest*, and *earnings* is a plural noun form of the verb *earn*.

Reading Comprehension

Words in Context Remind students that the same word can be used as different parts of speech, depending on the context. On the board, write, *Last year, it cost $1,890 to feed 540 runners*. Below it, write, *About how much money will be needed this year to cover the cost of snacks?* Guide students to recognize that the word *cost* is used as a verb in the first sentence and as a noun in the second sentence. As students read the problems in the unit, challenge them to find the word *plan* used as both a noun and a verb, and the word *total* used as a verb, an adjective, and a noun.

Listening and Speaking

Annually, Monthly, Weekly Tell students that *annually* is a synonym for *yearly*, and that both words are used to describe something that happens or is done once a year. Similarly, the word *weekly* describes something that happens or is done once a week, and the word *daily* describes something that happens or is done once a day. Have pairs of students take turns saying sentences to each other that contain the words *annually, monthly,* or *weekly*. For example, one student might say to the other, *I go to the dentist annually*.

Lesson 5: Strategy Focus — Make a Table

Lesson Overview

Lesson Materials: calculator

Skills to Know	Outcome	Math Vocabulary	eResources www.optionspublishing.com
• Use ratios and rates • Solve proportions	Students will recognize that tables are an efficient way to organize information in multi-step problems.	profit, rate, ratio	• Interactive Whiteboard Transparency 5 • Homework, Unit 2 Lesson 5 • Know-Find Table • Problem-Solving Checklist, also available in the student worktext, page 7

Modeled Instruction

Learn

To be sure students understand the context of the problem, ask questions such as the ones below.

- What are the profits of a business?
- Will the three people earn the same amount in profits? How do you know?

As students read the problem again, ask questions to help them focus on the details needed to solve it.

- How would you list the investors in order from the person who invested the most to the person who invested the least?
- What does $44,000 represent in the problem?

 Use a Graphic Organizer You may wish to use this Think Aloud to demonstrate how a graphic organizer is used to identify important information.

This problem provides information about three people's investments. I will use a Know-Find Table to see how I can use the information. In the Know column, I can write Ms. Jacobs invested $3,000, Mr. Kang invested $4,000, and Ms. Lopez invested $5,000. *I can also write* They will share the profits in the same ratio as they invested *and* The total profits are $44,000. *In the Find column, I can write* The amount of profit each person will receive.

I think the phrase they will share the profits in the same ratio as their investment *is important in solving the problem. But which ratios do I need to find? How will I keep the numbers straight? I think I need to find a way to organize the information so that I can see the relationships.*

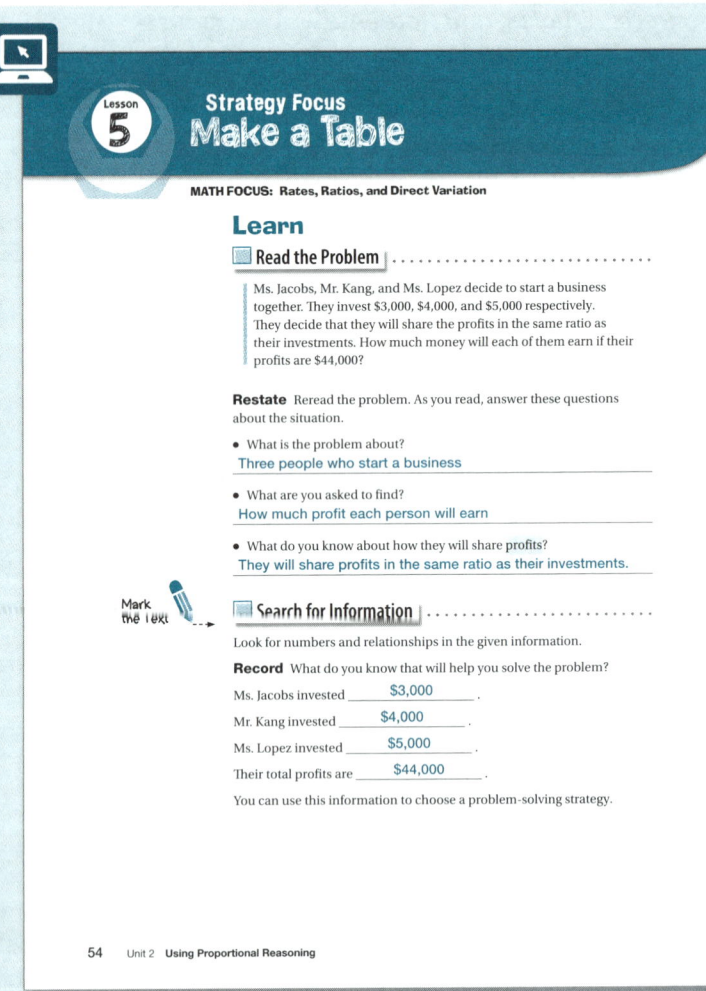

34   Unit 2

Decide What to Do

You know 3 investment amounts and the total profit.

Ask How can I find how much profit each investor earned?

- I can use the strategy *Make a Table*. I can record the initial investment for each person. I can find the sum of their investments.
- Then I can find how many times greater their profit is than their investment. Finally, I can find the amount earned by each investor.

Use Your Ideas

Step 1 Find the total investment.

	Investment	Multiplier	Profit
Ms. Jacobs	$3,000	$3\frac{2}{3}$	$11,000
Mr. Kang	$4,000	$3\frac{2}{3}$	$14,666.67
Ms. Lopez	$5,000	$3\frac{2}{3}$	$18,333.33
Total	$12,000	$3\frac{2}{3}$	$44,000

Step 2 Divide the profit by the total amount invested. This is your multiplier.

$44,000 \div \underline{\$12,000} = \underline{3\frac{2}{3}}$

Step 3 Since they share the profits in the same ratio as their investment, use the multiplier to find how much profit each investor earned. Round to the nearest cent.

$3,000 \times \underline{3\frac{2}{3}} = \underline{\$11,000}$
$4,000 \times \underline{3\frac{2}{3}} = \underline{\$14,666.67}$
$5,000 \times \underline{3\frac{2}{3}} = \underline{\$18,333.33}$

So Ms. Jacobs earned $\underline{\$11,000}$, Mr. Kang earned $\underline{\$14,666.67}$ and Ms. Lopez earned $\underline{\$18,333.33}$.

The multiplier is used because it keeps the shares of the profit in the same ratio to the total as the investment.

Review Your Work

Check that the sum of the individual profits is equal to the total profit.

Explain Dawn says the three investors should earn $13,666.67, $14,666.66, and $15,666.67. Do you agree with Dawn? Explain.

Sample Answer: No, because they should share the profits in the same ratio as their investments.

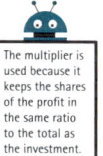

55

Modeled Instruction (continued)

Help students recognize how the details they have identified can be used to determine a problem-solving strategy.

- How many different answers will you find?
- How will a table help you organize the details?

Pose questions that help students focus on the steps used to solve the problem.

- How do you find the total amount invested?
- What is the ratio of the total profit to the total investment? Why is this ratio important?

Emphasize the importance of understanding that the three investors shared the entire $44,000 profit.

HOTS Explain Responses should note that the three investors' earnings should be related by proportions, not by addition.

Try

Solve the problem.

1 Juan earned $99.00 in overtime pay last week. When he works more than 8 hours in one day, Juan earns 1.5 times his regular hourly rate for the overtime hours. Last week, Juan worked 8 hours on Monday, 8 on Tuesday, 10 on Wednesday, 9 on Thursday, and 11 on Friday. What was Juan's total pay last week?

Read the Problem and Search for Information

Mark the numbers and words that will help you solve the problem.

Decide What to Do and Use Your Ideas

You can use the strategy *Make a Table* to organize the information.

Step 1 Record the pay at the overtime rate, and the hours at each rate.

	Hours	Hourly Rate	Pay
Regular	40	$11.00	$440.00
Overtime	6	$16.50	$99.00

Ask Yourself: How many hours did Juan work in all? How many hours were overtime?

Step 2 Find the hourly rate for overtime hours using the table.

Step 3 Find the hourly rate for regular hours.

Regular rate × 1.5 = $\underline{\$16.50}$
Regular rate = $\underline{\$11.00}$

Step 4 Find the total pay for the week.

Regular Pay = 40 × $\underline{\$11.00}$
Regular Pay + Overtime Pay = Total Pay
$\underline{\$440.00} + \$99.00 = \underline{\$539.00}$

So Juan's total pay last week was $\underline{\$539.00}$.

Review Your Work

Check that your calculations are correct.

Recognize What led you to make a table to solve this problem?

Sample Answer: There were two rates of pay for different numbers of hours, so a table helps me keep the numbers organized.

56 Unit 2 Using Proportional Reasoning

Guided Practice

Try

1 Remind students that a table can help them see what numbers are needed to find the solution.

Discuss the concept of overtime in the problem.

- How is Juan's overtime rate calculated?
- Why might someone be paid extra to work overtime?

Guide students to begin with what they know and to then think through the steps needed to solve the problem.

- Why would you begin by finding the number of overtime hours that Juan worked?
- How do you find Juan's regular hourly rate?

Have students explain what operations they can use to check that their calculations are correct.

HOTS Recognize Students' explanations should discuss the need to keep the hours, rates, and pay organized.

Lesson 5 35

Scaffolded Practice
Apply

(2) Help students think through the steps needed to solve the problem.

- *What steps will you follow to find the estimated monthly cost of the new plan?*

HOTS Apply Explanations should describe how to find the difference in the fixed monthly fees and then divide that difference by 0.08 to find the additional minutes used.

(3) Prompt students to recognize what information they need to know in order to solve the problem.

- *What do you need to know to find each student's total allowance for the school year?*
- *Why do you need to find the total of the students' allowances for one week to find the number of weeks?*

HOTS Choose Responses should show that multiplication is used when an amount is added repeatedly.

(4) Make sure students understand the problem context.

- *What does* miles per gallon *mean?*
- *What operation do you use to find the number of miles per gallon?*

HOTS Explain Responses should note how a table organizes numbers and makes it easier to record and compare computations.

(5) Have students consider the steps needed to solve the problem.

- *How can you find the amount spent on each runner for safety services? For snacks?*
- *Once you know the costs per runner, how do you find the total costs?*

HOTS Decide Explanations should describe how to label the rows with the types of expenses and the columns with the rates, numbers, and costs needed to solve the problem.

Apply

Solve the problems.

(2) Lee is looking for a new cell phone plan. During the last 6 months, Lee used an average of 595 minutes each month. His current plan is $59.99 per month with unlimited minutes. He is considering switching to a plan that costs $49.99 per month with 500 minutes included, plus 8 cents for each additional minute. Which plan will cost less? Explain.

Ask Yourself: How can I determine the monthly cost for the new plan?

Plan	Monthly Fee	Minutes Included	Cost per Additional Minute	Additional Minutes	Estimated Monthly Cost
Current	$59.99	Unlimited	$0.00	0	$59.99
New	$49.99	500	$0.08	about 95	$57.59

Hint: Use the table to help you organize the information in the problem.

Answer: The new plan will cost less. Lee will save about $2.40 each month.

Apply How many minutes would Lee have to use each month for the two plans to cost the same? Explain your reasoning.

Sample Answer: Lee would have to use 625 minutes each month because $59.99 = 0.08(x - 500) + 49.99$.

(3) Nora, Opal, Paulo, and Quinn have weekly allowances of $2, $3, $4, and $5 respectively. During the school year, their allowances total $546. How much did each of them receive during the school year?

Ask Yourself: How can I find the number of weeks they receive allowances during the school year?

	Allowance	Number of Weeks	Total Allowance
Nora	$2	39	$78
Opal	$3	39	$117
Paulo	$4	39	$156
Quinn	$5	39	$195
Total	$14	39	$546

Hint: Find the total of the weekly allowances first.

Answer: The amounts they received were Nora $78, Opal $117, Paulo $156, and Quinn $195.

Choose What operation did you use to find the total for each person?

Sample Answer: I multiplied each weekly allowance by the number of weeks because the allowances are the same each week.

Lesson 5 Strategy Focus: Make a Table 57

---

(4) During their vacation, the Chan family recorded how far they had traveled each time they stopped for gas. Each time they stopped for gas, they filled the gas tank. They stopped after 380 miles and bought 13.5 gallons of gas. After another 402 miles they bought 11.6 gallons of gas. They stopped a last time after another 299 miles and bought 9.6 gallons of gas. During which part of the trip did the car get the greatest number of miles per gallon?

Ask Yourself: How do I find the miles per gallon?

Hint: Round the miles per gallon to the nearest tenth.

Fill-up	Miles Since Last Fill-up	Number of Gallons of Gas Used	Approximate Miles per Gallon
1	380	13.5	28.1
2	402	11.6	34.6
3	299	9.6	31.1

Answer: The car got the greatest number of miles per gallon during the second part of the trip.

Explain How did using a table help you solve this problem?

Sample Answer: A table helped me organize the information so that the numbers I needed to compare were in the last column.

(5) The organizers of a road race use donations to provide snacks and safety services for the runners. This year and last year, safety services cost $50 for every 100 runners. Last year, it cost $1,890 to feed 540 runners. This year, they expect 120 more runners than last year. About how much money will be needed this year to cover the cost of snacks? How much money will be needed to provide safety services?

Ask Yourself: How should I label the column and rows?

Hint: Find the cost per runner.

	Rate	Rate per Runner	Number of Runners	Total Cost
Safety Services	$50 for 100 runners	$0.50	660	$330
Snacks	$1,890 for 540 runners	$3.50	660	$2,310

Answer: About $2,310 will be needed for snacks and $330 will be needed for safety services.

Decide How did you decide what headings to use for the table?

Sample Answer: I put the two expenses in the rows. I included the rates as given, the rate per runner, and the number of runners as columns so that I could find the total cost.

Practice

Solve the problems. Show your work.

6. Doreen makes pizza for a local pizza restaurant. She worked 28 hours last week and earned $313.60. She worked 18 hours this week. How much money will Doreen earn for her work this week?

Some students will use the lesson strategy; however, other strategies may be used. Accept all reasonable work leading to the correct answer.

Answer Doreen will earn $201.60 this week.

Consider Why is it important to go back and reread a problem after solving it?

Sample Answer: Rereading a problem helps me be sure that I have answered the question that was asked.

7. Members of a swim team sold T-shirts and sweatshirts with the team logo to raise money for their swim meets. They made a profit of $4 on each T-shirt and $7.50 on each sweatshirt. They made a total profit of $437. If they sold the same number of T-shirts as sweatshirts, how much profit did they make from each?

Some students will use the lesson strategy; however, other strategies may be used. Accept all reasonable work leading to the correct answer.

Answer They made $152 from selling T-shirts and $285 from selling sweatshirts.

Compare How is this problem similar to Problem 3? How is it different?

Sample Answer: Both problems involve distributing dollar amounts using a ratio. Problem 3 has a ratio with 4 parts. This problem has a ratio with 2 parts.

Create Look back at Problem 5. Change the kinds of expenses and the number of runners. Write and solve a problem about organizing a road race. Be sure you can solve your problem using the strategy *Make a Table*. See teacher notes.

Lesson 5 Strategy Focus: Make a Table   59

Create

In this lesson, students modify Problem 5 by changing two pieces of information, the number of runners, and the types of expenses. If students are struggling, suggest they change the costs of expenses and the number of runners to compatible numbers.

Accept student responses that that make the requested changes, organize the information using a table, and provide a correct solution.

Independent Practice
Practice

Students should be encouraged to choose any strategy to solve Problems 6 and 7, though many may prefer to use *Make a Table*.

6. Some students may divide last week's pay by the number of hours worked to find the hourly rate and then multiply to find this week's pay, without making a table. If students choose to make a table, guide them to include a column labeled *Rate of Pay*.

Sample Work

Week	Hours Worked	Rate of Pay	Total Earned
1	28	$11.20	$313.60
2	18	$11.20	$201.60

HOTS Consider Students' responses should note that it is important to reread a problem to see if they answered the correct question.

7. Guide students to understand that they can divide the total profit by the sum of the profit on each T-shirt and sweatshirt to find the total number of each sold.

Sample Work

Item	Number	Profit on Each	Total Profit
T-shirt	38	$4.00	$152.00
Sweatshirt	38	$7.50	$285.00
Total	38	$11.50	$437.00

HOTS Compare Students' explanations should show they understand that the problems can be solved in a similar manner, but that Problem 3 has more parts to the ratio.

Lesson 5   37

Lesson 6
Strategy Focus: Write an Equation

Lesson Overview

Lesson Materials: calculator

Skills to Know	Outcome	Math Vocabulary	eResources www.optionspublishing.com
• Solve percent equations and proportions • Interpret circle graphs	Students will recognize that writing equations is an efficient way to solve problems involving percent.	budget, percent equation, proportion	• Interactive Whiteboard Transparency 6 • Homework, Unit 2 Lesson 6 • Problem-Solving Checklist, also available in the student worktext, page 7

Modeled Instruction
Learn

Probe students' understanding of the problem's context by asking questions similar to the following.

- *What are the categories for the speech club's expenses listed in order from greatest percent to least percent?*
- *Which expense is the smallest portion of the budget? How did you decide?*

As students read the problem again, guide them to identify the words and numbers needed to solve the problem.

- *What amount of money does the entire circle graph represent? How do you know?*

Reread You may wish to use this Think Aloud to demonstrate how to read a problem to recognize important information.

I know this problem is about a speech club going on a trip. They have a budget of $3,750 to spend on all of the costs of the trip. I need to find how much of that money they can spend on tickets to a museum.

I see there is a circle graph. It shows how the club plans to spend the $3,750. It looks like most of the money will be spent on food, lodging, and transportation. It looks like a smaller portion of the money will go to tickets.

I see that the graph gives percents for every category except tickets. I wonder if there is a way to find the percent of the money spent on tickets. That still does not tell me how much money they can spend on tickets. But maybe it will help me get started.

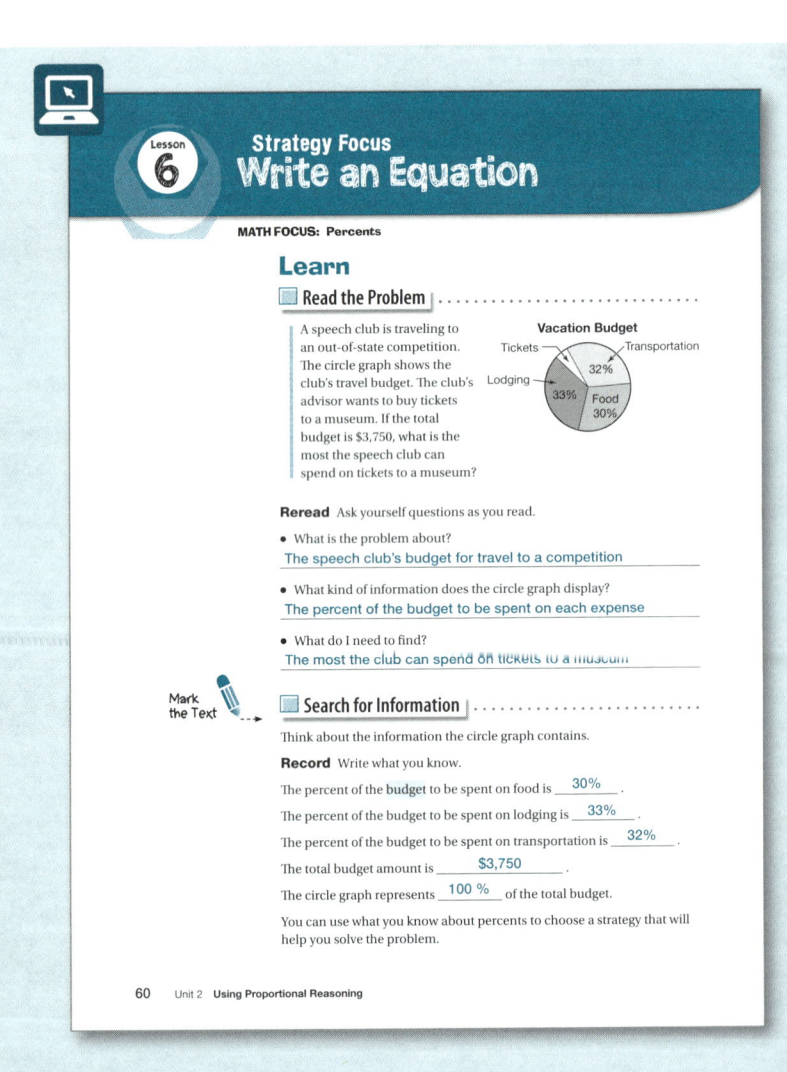

38 Unit 2

Decide What to Do

You know the percents the club plans to spend on food, lodging, and transportation. You also know the total budget.

Ask How can I find the amount the speech club can spend on tickets?

- I can find the percent of the budget left for tickets.
- Then I can use the strategy *Write an Equation* to find the amount the club can spend on tickets.

Use Your Ideas

Step 1 Find the percent of the budget left for tickets.

Percent of budget for other categories:
30% + 33% + 32% = __95%__
Percent of budget for tickets: 100% − 95% = __5%__

Step 2 Find the amount the club can spend on tickets using the **percent equation**. Write a word equation first. Let t represent the amount the club can spend on tickets, in dollars.

part = percent × whole
$t =$ __0.05__ × $3,750 ← Write the percent as a decimal.
$t =$ __$187.50__ ← Simplify.

You can write 5% as the decimal 0.05.

The most the club can spend on tickets to a museum is __$187.50__.

Review Your Work

Make sure that you answered the question asked. The question asks for the amount, not the percent.

Identify For this problem, what do the words *part*, *percent*, and *whole* represent in the word equation?

Sample Answer: Whole represents the total budget. Percent represents the percent that can be spent on tickets. Part represents the dollar amount that can be spent on tickets.

61

Try

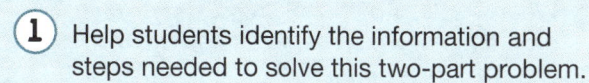

Solve the problem.

1. The athletic department is planning an awards ceremony. Teams are helping to raise money. So far, the teams have raised $1,860. The athletic department will spend $1,200, or 48% of the total budget, on trophies. What percent of the total budget must still be raised?

Read the Problem and Search for Information

Think about whether the answer will be an amount or a percent.

Decide What to Do and Use Your Ideas

When you need to find a whole or a percent, you can write a proportion.

Step 1 Find the total budget, b. Write and solve a proportion.

Ask Yourself: How do I write a percent as a fraction?

$$\frac{part}{whole} = \frac{\%}{100}$$

$$\frac{\$1,200}{b} = \frac{48}{100}$$

$$100 \times \underline{\$1,200} = \underline{48} \times b$$

$$\underline{\$2,500} = b$$

Step 2 Find the amount of money that must still be raised.

Total budget − Amount raised = Amount to be raised
__$2,500__ − $1,860 = __$640__

Step 3 Find the percent, x, that must still be raised. Write and solve a proportion.

$$\frac{}{whole} = \frac{}{100}$$

$$\frac{\$640}{\$2,500} = \frac{x}{100}$$

$$25.6 = x$$

So __25.6%__ of the total budget must still be raised.

Review Your Work

Did you substitute the correct values into the equations?

Explain How can you use estimation to check your work?
Sample Answer: 48% is almost 50%. Since $1,200 is about 50% of the budget, the budget is about $2,400. The teams have already raised about $\frac{1,800}{2,400}$, which is $\frac{3}{4}$ (75%), of the total budget. So about 25% must still be raised.

62 Unit 2 Using Proportional Reasoning

Scaffolded Practice
Apply

2 Help students differentiate between the information that represents a part and that which represents a whole.

- Does the total cost of the hotel rooms represent the part or the whole in the percent proportion? How did you decide?

HOTS Determine Explanations should note that it is not necessary to know that four students share a room.

3 Guide students through the steps they will use to solve the problem.

- What equation can you use to find the total cost?
- Once you know the total cost, what do you do next?

HOTS Examine Responses should state that Jane only included one month's worth of food in the total.

4 Have students consider how the information provided relates to the solution.

- What words or numbers could you mark in the problem that would help you find the monthly entertainment budget?
- What information can you determine from the graph? How will you use this information?

HOTS Analyze Students' responses should show that they found the percent spent on entertainment using the percents on the circle graph, and then they used that percent to find the total budget.

5 Help students focus on the information needed to solve the problem.

- What numbers from the problem can you use to find the total Claudia earned last year? This year?

HOTS Verify Students' explanations should indicate that the first proportion helps them determine Claudia's earnings for last year and the second helps them determine the earnings needed to reach her goal this year.

Apply
Solve the problems.

2 Forty-four students are going on a class trip for 2 days and 1 night. If 4 students sleep in each room, the cost per student will be $60 per night. Two chaperones are also going. Their rooms are each $225 per night. The total budget for the trip is $6,000. What percent of the budget will be spent on hotel rooms?

Hint: First find the total cost for all the hotel rooms.

Total cost of rooms = 44 × __$60__ + 2 × __$225__
= __$3,090__

$\dfrac{\text{part}}{\text{whole}} = \dfrac{\%}{100}$

$\dfrac{\$3,090}{\$6,000} = \dfrac{\%}{100}$

Ask Yourself: How much will the rooms for the students cost? How much will the rooms for the chaperones cost?

Answer __51.5% of the budget will be spent on hotel rooms.__

Determine What information is given that is not needed?

Sample Answer: You don't need to know that 4 students sleep in each room.

3 José wants to adopt a dog. He must earn enough money to pay for the dog and all expenses for one year. It costs $250 to adopt a dog and pay for all of its shots. Food costs $20 per month. Other supplies cost $120 annually. José has already earned 70% of the total cost for the first year. How much more must he earn before he can adopt a dog?

Total costs = __$610__
Percent left to earn = __30%__
Part = percent × whole
a = __0.3__ × __$610__

Ask Yourself: What percent of the total cost must José still earn?

Hint: Write an equation using the total cost and the percent he has left to earn.

Answer __José must earn $183 more before he can adopt a dog.__

Examine Jane says that José must still earn $117. What mistake could she have made?

Sample Answer: Jane only counted $20 once. The food is a monthly cost, so she needs to multiply $20 by 12.

Lesson 6 Strategy Focus: Write an Equation 63

---

4 The circle graph shows how the Teen Drop-In Center usually budgets its money each month. The center's staff is planning to buy a digital piano. They will save $\frac{2}{3}$ of the usual entertainment budget each month for four months. If the staff saves $320 each month for the piano, what is the center's total monthly budget?

Monthly Budget
Entertainment — Food/Necessities 45%
Utilities 15%
Housing 30%

Ask Yourself: What percent of the monthly budget does the center's staff usually spend on entertainment?

Hint: Write a proportion. Let b be the total monthly entertainment budget, in dollars.

$\dfrac{\$320}{b} = \dfrac{2}{3}$

Answer __The center's total monthly budget is $4,800.__

Analyze After you solved the proportion, there was still another step needed. Describe how you finished solving the problem.

Sample Answer: I knew the entertainment budget was 10% of the entire monthly budget, so I multiplied b by 10 to answer the question.

5 Claudia saved $141 last year. That was 30% of the money she earned by babysitting. She plans to continue saving 30% of her babysitting money this year. She hopes to save $180. How much more money must Claudia earn this year than last year to meet her goal?

Ask Yourself: How can I use the amounts Claudia saves to find the amounts she earns?

Last year: $\dfrac{\$141}{m} = \dfrac{30}{100}$

Hint: Claudia's savings is the part in each equation.

This year: $\dfrac{\$180}{n} = \dfrac{30}{100}$

Answer __Claudia must earn $130 more this year than last year.__

Verify Why did you need to use two proportions to solve the problem?

Sample Answer: I needed to find the total amount Claudia earned last year and the total amount she would need to earn this year.

Unit 2 Using Proportional Reasoning

Practice

Solve the problems. Show your work.

6. Students in Hill Lane Middle School are selling T-shirts for $13 each during a fund raiser. So far, they have raised 65% of their goal of $5,000. How many T-shirts have they sold so far?

Some students will use the lesson strategy; however, other strategies may be used. Accept all reasonable work leading to the correct answer.

Answer They have sold 250 T-shirts so far.

Formulate What is another question you could ask from the information given in the problem?

Sample Answer: How many more T-shirts must the students sell to meet their goal of $5,000?

7. Ryan wants to get an after-school job. His parents will let him work only if he can still spend 15 hours on school work during the week. He makes this circle graph to show how he can do everything he needs to do after school, including the job and 15 hours of school work. How many hours does Ryan plan to work during the week?

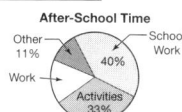

After-School Time
Other 11%
School Work 40%
Work
Activities 33%

Some students will use the lesson strategy; however, other strategies may be used. Accept all reasonable work leading to the correct answer.

Answer Ryan plans to work 6 hours during the week.

Discuss How did you use equations to solve the problem?

Sample Answer: I used 15 hours and 40% to write a proportion to find that the total number of after-school hours was 37.5. Then I found that 16% of the time was left for work and wrote an equation to find that 16% of 37.5 is 6.

Create Look back at Problem 3. Write a new problem by changing at least two of the costs in the problem and the percent that José has saved. Solve your problem. See teacher notes.

Lesson 6 Strategy Focus: Write an Equation   65

Create

In this lesson, students modify Problem 3 by changing two of the costs of owning a dog as well as changing the percent José has saved. If students are struggling, suggest they use multiples of $100 for the cost and a percent that is easy to work with, such as 25%, 50%, or 75%.

Accept student responses that change the three quantities and provide a correct solution. Students should be able to use an equation to solve the problem.

Independent Practice
Practice

Students should be encouraged to choose any strategy to solve Problems 6 and 7, though many may prefer to use *Write an Equation*.

6. Some students may choose to first find the amount raised and then the number of T-shirts, while others may write just one equation. Be sure students find the number of T-shirts sold, not the amount of money raised selling T-shirts.

Sample Work

Let t = the number of shirts sold.

$13t = 0.65 \times 5,000$

$13t = 3,250$

$t = 250$

HOTS Formulate Responses should show that students understand how the information in the problem can be used to ask another question.

7. Remind students to read the problem carefully so that they answer the correct question.

Sample Work

Let t = total free time outside school and w = percent of time allotted to work.

$\dfrac{15}{t} = \dfrac{40}{100}$

$40t = 1,500$

$t = 37.5$

$100\% = (11\% + 40\% + 33\%) + w$

$100\% = 84\% + w$

$16\% = w$

Let h = number of work hours.

$h = 16\% \times 37.5$

$h = 0.16(37.5)$

$h = 6$

HOTS Discuss Students' explanations should describe using a proportion to find the total hours Ryan has outside of school and equations to find both the percent and the amount of time he can spend at work.

Lesson 6   41

Lesson 7

Strategy Focus: Work Backward

Lesson Overview

Lesson Materials: calculator

Skills to Know	Outcome	Math Vocabulary	eResources www.optionspublishing.com
• Solve percent problems	Students will recognize that working backward is an efficient way to find an initial amount when a final amount is given.	interest, tax	• Interactive Whiteboard Transparency 7 • Homework, Unit 7 Lesson 7 • Know-Find Table • Problem-Solving Checklist, also available in the student worktext, page 7

Modeled Instruction

Learn

Probe students' understanding of the problem's context by asking questions similar to the following.

- *Where did Tara get the money for her shopping trip?*
- *Besides paying the prices of the items she bought, what else did Tara have to pay?*

As students read the problem again, ask questions to help them focus on the details needed to solve it.

- *Which numbers would you use to find the cost of the items Tara purchased?*
- *How is the number 6% different from the other numbers in the problem?*

Use a Graphic Organizer You may wish to use this Think Aloud to demonstrate how to record and sort the information needed to solve the problem.

This problem gives a lot of numbers about a shopping trip. I will use a Know-Find Table to help me see what I know and figure out what I need to find. In the Know column, I will put the cost of each item. I can also put in the Know column that The sales tax is 6%, *and that* Tara had $5.50 left in change. *In the Find Column, I can put* How much money Tara earned last month *and* The total amount Tara spent.

I know the amount of money Tara had at the end of her shopping trip. I need to find how much she earned last month. I think I need to find how much Tara spent including the sales tax. I wonder how I can find that amount?

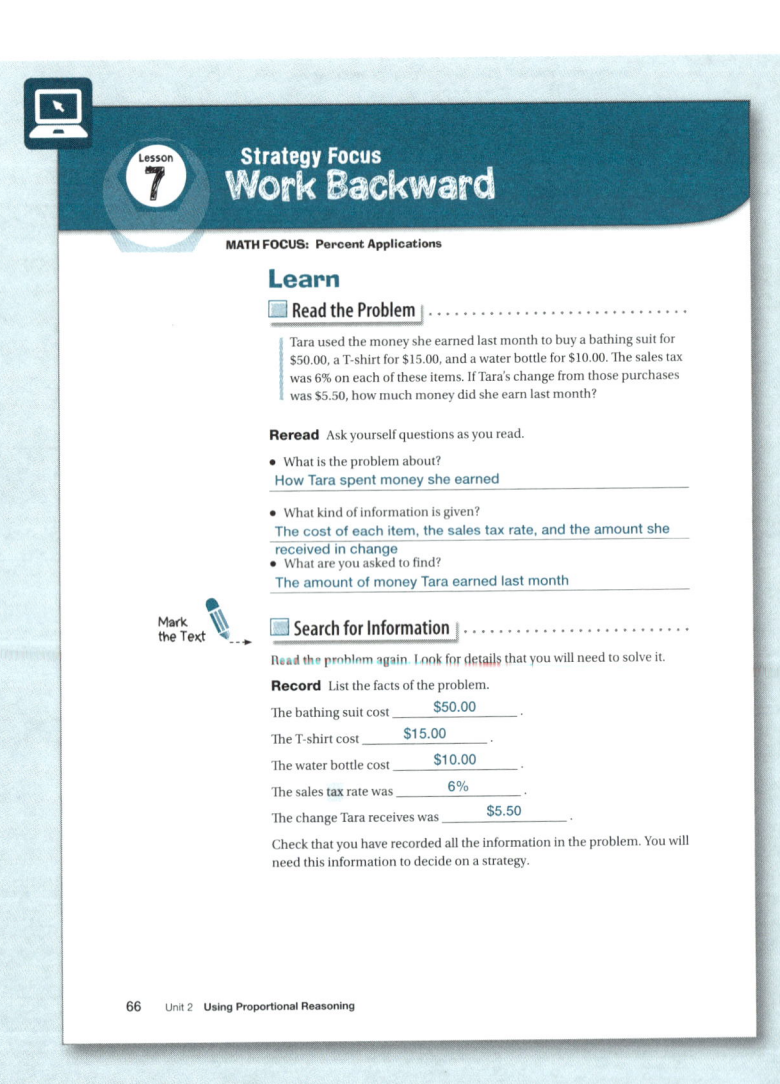

Lesson 7 — Strategy Focus: Work Backward

MATH FOCUS: Percent Applications

Learn

Read the Problem

Tara used the money she earned last month to buy a bathing suit for $50.00, a T-shirt for $15.00, and a water bottle for $10.00. The sales tax was 6% on each of these items. If Tara's change from those purchases was $5.50, how much money did she earn last month?

Reread Ask yourself questions as you read.

- What is the problem about?
  How Tara spent money she earned
- What kind of information is given?
  The cost of each item, the sales tax rate, and the amount she received in change
- What are you asked to find?
  The amount of money Tara earned last month

Mark the Text

Search for Information

Read the problem again. Look for details that you will need to solve it.

Record List the facts of the problem.

The bathing suit cost _____$50.00_____.
The T-shirt cost _____$15.00_____.
The water bottle cost _____$10.00_____.
The sales tax rate was _____6%_____.
The change Tara receives was _____$5.50_____.

Check that you have recorded all the information in the problem. You will need this information to decide on a strategy.

66   Unit 2   Using Proportional Reasoning

---

42   Unit 2

Decide What to Do

You know the amount of money that Tara received in change. You know how much she spent and what the sales tax rate was.

Ask How can I find the amount Tara earned last month?
- I can use the strategy *Work Backward*.
- I can add back the amount Tara spent to the amount she got in change to find how much Tara earned.

You know an ending amount. You need to find a starting amount.

Use Your Ideas

Step 1 Find the total cost of the items Tara bought.

$50.00 + $15.00 + $10.00 = __$75.00__.

Step 2 Find the amount of the sales tax on that total.

__$75.00__ × 0.06 = __$4.50__

Step 3 Add the amount Tara spent to the change she received.

Cost of items + Tax + Change = Amount Tara earned

__$75.00__ + __$4.50__ + $5.50 = __$85.00__

So Tara earned __$85.00__ last month.

Review Your Work

Check that you have answered the question that was asked and that your answer makes sense in the problem.

Identify How do you know that your answer makes sense?

Sample Answer: I know that the total cost of the items was $75.00, so I know that my answer must be greater than that amount.

Try

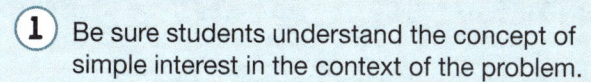

Solve the problem.

1. Two years ago, Karl invested some money in an account that pays 4% simple interest annually. He just withdrew 45% of the interest that his money has earned. If Karl withdrew $144, how much money did he invest in the account two years ago?

Read the Problem and Search for Information

Reread the problem and mark the information that will help you.

Decide What to Do and Use Your Ideas

You can use the strategy *Work Backward* to find the total amount of interest Karl earned. Then you can find the amount that Karl invested.

Step 1 Let I be the amount of interest. Write an equation to find the total amount of interest earned in two years.

Amount Karl withdrew = __$144__
Percent of total interest withdrawn = __45%__
Percent × Interest earned = Part
$0.45 \times I =$ __144__
$I = \frac{144}{0.45}$
$I =$ __320__

Step 2 Find the amount that Karl invested. Let P represent the principal, the amount he invested. Write the interest rate as a decimal.

Principal × Rate × Time = Interest
$P \times$ __0.04__ × 2 = __320__
$P =$ __4,000__

Two years ago, Karl invested __$4,000__.

Ask Yourself: How do I write 4% as a decimal?

Review Your Work

To check your work, start with your answer and work forward to the time when Karl withdrew the money.

Generalize Why do you need to use inverse operations when you work backward to solve a problem?

Sample Answer: To get back to the starting amount, you need to undo the actions in the problem.

Scaffolded Practice
Apply

2 Prompt students to consider how to think about the information in order to solve it.
- What words or numbers in the problem help you find the percent of pay Ms. Marino takes home? Why is this important to know?
- What information helps you determine the number of weekly paychecks Ms. Marino received?

HOTS Explain Responses should point out that the problem gives a final amount and asks for a beginning amount.

3 Have students think about how they can use percents to solve the problem.
- Why can you think about the total amount paid as 105% of the sale price?
- What percent of the original price is the sale price? How does this help you?

HOTS Judge Responses should cite words such as *original* that suggest working backward might be a good strategy.

4 Help students think about the steps needed to solve the problem.
- What will you do first? What is the next step?
- How can you use the total sales and total profit to find the amount paid for the raincoats?

HOTS Justify Explanations should note that slightly less than half of the selling price is profit, so the amount paid would be slightly more than half of $34.50.

5 Guide students to figure out how to work backward from the take-home pay to find the total sales.
- What numbers do you start with to find Ms. Newton's earnings before taxes and insurance? Why?

HOTS Contrast Explanations should note that both problems are about pay after deductions, but this problem involves commission.

Apply
Solve the problems.

2 Last year, Ms. Marino had 28% of her earnings deducted from her paycheck each week. She took 2 weeks off from work without pay. If she took home a weekly paycheck of $306, how much did Ms. Marino earn before deductions last year?

Ask Yourself: If 28% of Ms. Marino's earnings were deducted from her paycheck, what percent of her earnings were left?

Earnings before deductions × Percent left = Take home pay

$e \times$ __0.72__ = $306

$e =$ __$425__

Total earnings = Take home pay × Number of weeks worked

= __425__ × __50__

Hint: There are 52 weeks in one year.

Total earnings = __$21,250__

Answer Last year, Ms. Marino earned $21,250 before deductions.

Explain Why could you use the *Work Backward* strategy to solve this problem?

Sample Answer: I knew the ending amount, which was Ms. Marino's take-home pay. I needed to find the beginning amount, which was Ms. Marino's earnings before deductions.

3 Mr. Suarez bought a new computer. It was on sale for 20% off the original price. Including a 5% sales tax, Mr. Suarez paid $587.16 for the computer. What was the original price?

Ask Yourself: How can I find the price before the tax and the sale?

Sale Price × Percent = Total paid

$s \times 1.05 =$ __$587.16__

$s =$ __$559.20__

Hint: Think of the total paid as 105% of the sale price. Think of the sale price as 80% of the original price.

Original price × Percent of sale price = Sale price

$P \times 0.8 =$ __$559.20__

$P =$ __$699.00__

Answer The original price was $699.00.

Judge What words in a problem suggest that working backward might be a good strategy?

Sample Answer: Words like *before*, *start*, and *original* suggest that working backward will be helpful in solving the problem.

---

4 To raise money for school sports, the sports teams decided to sell raincoats in school colors. They bought 12 dozen raincoats. They sold the raincoats for $34.50 each. The profit was 45% of the total sales. How much did the sports teams pay for each raincoat?

Ask Yourself: How many raincoats were sold?

The total sales amount was __$4,968__.

The profit on the sales was __$2,235.60__.

The total amount paid for the raincoats was __$2,732.40__.

Hint: Round your answer to the nearest cent.

Answer The school paid $18.98 for each raincoat.

Justify How can you estimate to be sure that your answer makes sense?

Sample Answer: The profit was a little less than half of the total sales. The students must have paid a little more than half of what they sold the raincoats for.

5 Ms. Newton gets paid a weekly base salary of $240. She also earns 4% commission on her sales of mobile phones and accessories. She has 25% of her total earnings deducted for taxes and health insurance. Her take home pay last week was $420. What were Ms. Newton's total sales for last week?

Hint: If 25% of her pay is deducted, then 75% is left.

Ask Yourself: What equation can I write to find Ms. Newton's earnings before taxes and insurance are deducted?

Earnings before taxes and insurance were __$560__.

Pay from commissions was __$320__.

Answer Ms. Newton had total sales of $8,000 last week.

Contrast How is this problem different from Problem 2?

Sample Answer: They both are about take home pay, but this problem involved commissions.

Practice

Solve the problems. Show your work.

6 A group of friends pooled their money to go out for pizza. A tax of 7% was added to the cost of the meal. The friends left a tip of $6.50, which was 20% of the cost of the meal before the tax was added. What was the cost of the meal before tax and tip?

Some students will use the lesson strategy; however other strategies may be used. Accept all reasonable work leading to the correct answer.

Answer The cost of the meal before tax and tip was $32.50.

Conclude What information is given that is not necessary to solve this problem?
Sample Answer: The tax rate of 7% is not needed to solve this problem.

7 Two months ago, Chico opened a savings account that pays 0.70% interest at the end of every two months. He has neither withdrawn nor deposited any money since he opened the account. He now has $266.86 in the account. How much money did Chico deposit into the account when he opened it?

Some students will use the lesson strategy; however other strategies may be used. Accept all reasonable work leading to the correct answer.

Answer Chico deposited $265.00.

Compare Which other problem in this lesson is most like Problem 7? Explain.
Sample Answer: Problem 1 is like Problem 7 since both use simple interest after 2 units of time.

Create Look back at Problem 6. Change the amount of the tip and the percent for the tip. Write and solve a problem about a group of friends eating at a restaurant. See teacher notes.

Lesson 7 **Strategy Focus: Work Backward** 71

Create

In this lesson, students change two numbers in Problem 6. They should use their own numbers for the tip and the percent for the tip. If students are struggling, suggest they start with the bill total before the tip, have them choose a percent for the tip, and then find the tip amount. Then have students write the problem and show how it can be solved.

Accept student responses that include a problem with the requested changes and that give the correct answer.

Independent Practice
Practice

Students should be encouraged to choose any strategy to solve Problems 6 and 7, though many may prefer to use *Work Backward*.

6 Some students may choose to write equations and solve the problem directly.

Sample Work

20% of the meal = tip

$$0.2m = 6.50$$
$$m = 6.50 \div 0.2$$
$$m = 32.50$$

Conclude Students' responses should state that they did not need the tax rate to find the solution.

7 Suggest students work the problem forward using variables for the unknown quantities, then work backward to find the solution.

Sample Work

Money invested × 1.007 = Money after 2 months

$$m \times 1.07 = \$266.86$$
$$m = \$266.86 \div 1.007$$
$$m = \$265.00$$

Compare Students' explanations should indicate that Problem 1 is also about interest.

Lesson 7   45

Lesson 8

Strategy Focus: Write an Equation

Lesson Overview

Lesson Materials: calculator

Skills to Know	Outcome	eResources www.optionspublishing.com
• Solve problems involving percent of change	Students will recognize that writing equations is an efficient way to represent real-world situations mathematically.	• Interactive Whiteboard Transparency 8 • Homework, Unit 2 Lesson 8 • Problem-Solving Checklist, also available in the student worktext, page 7

Modeled Instruction

Learn

To be sure students understand the context of the problem, ask questions such as the ones below.

- Why does North Middle School raise money every year?
- In what ways is this year's fundraiser different from last year's?

As students read the problem again, pose questions to help them recognize important phrases and facts.

- What words or numbers tell you how much the school wants to raise this year?

Reread You may wish to use this Think Aloud to demonstrate how to read a problem for different purposes.

The problem is about a school that raises money for a food pantry. Every year, they have a different type of fundraiser. I am going to read the problem again to see what I need to find. If I know what the problem is asking, maybe I will be able to figure out how to solve it.

I see that last year the school raised $1,500. This year they want to raise $20 more. Wait—that's not right. I misread the problem. They want to raise 20% more, not $20 more.

But what do I do now? Maybe I can multiply by 20%. But 20% of $1,500 is less than $1,500. I know the school wants to earn more money, not less money. I need to think a little more about the problem.

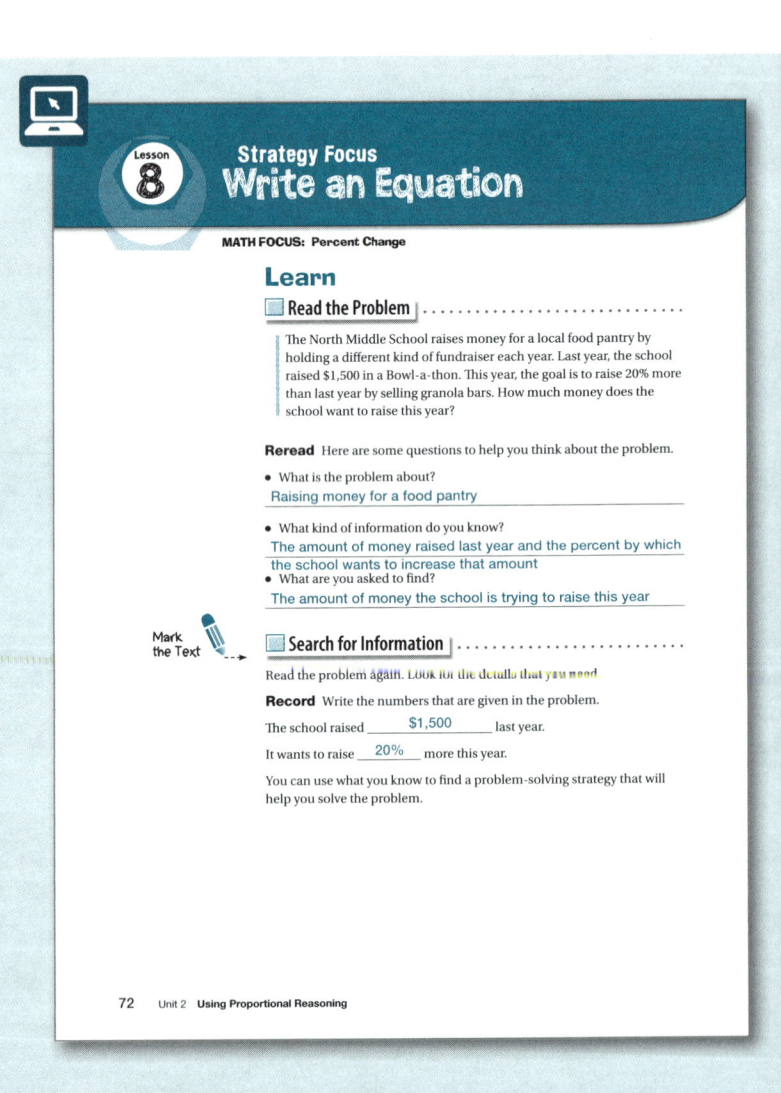

Lesson 8 — Strategy Focus: Write an Equation

MATH FOCUS: Percent Change

Learn

Read the Problem

The North Middle School raises money for a local food pantry by holding a different kind of fundraiser each year. Last year, the school raised $1,500 in a Bowl-a-thon. This year, the goal is to raise 20% more than last year by selling granola bars. How much money does the school want to raise this year?

Reread Here are some questions to help you think about the problem.

- What is the problem about?
  Raising money for a food pantry
- What kind of information do you know?
  The amount of money raised last year and the percent by which the school wants to increase that amount
- What are you asked to find?
  The amount of money the school is trying to raise this year

Mark the Text

Search for Information

Read the problem again. Look for the details that you need.

Record Write the numbers that are given in the problem.

The school raised ____$1,500____ last year.

It wants to raise ____20%____ more this year.

You can use what you know to find a problem-solving strategy that will help you solve the problem.

72   Unit 2   Using Proportional Reasoning

---

46   Unit 2

Decide What to Do

You know how much money was raised last year. You know the percent by which the school wants to increase that amount this year.

Ask How can I find the amount of money the school wants to raise this year?

- I can use the strategy *Write an Equation*.
- I can use the percent equation to find the amount of the increase. Then I can add the result to the amount raised last year.

Use Your Ideas

Step 1 Find the amount of increase. Use the percent equation.

part = percent × whole
Amount of Increase = 20% × $1,500
= 0.20 × $1,500
= __$300__

You can use an equation to represent the situation.

Step 2 Add the amount of increase to the amount raised last year.

Goal for this year = Amount last year + Amount of increase
= $1,500 + __$300__
= __$1,800__

So the school wants to raise __$1,800__ this year.

Review Your Work

Reread the problem to be sure that you have answered the question that was asked.

Describe How did writing an equation help you solve this problem?

Sample Answer: An equation helped me relate the facts I knew to each other. It reminded me how to use them to find what I didn't know.

73

Modeled Instruction *(continued)*

Help students make a connection between what they know and what they need to find out.

- *How does the percent equation help you find the amount of increase?*

Ask questions that encourage students to think critically about the steps in the solution process.

- *What part of the percent equation is unknown?*
- *Why would you add the amount of increase to last year's total?*

Emphasize the importance of answering the correct question by posing other questions about the problem.

HOTS Describe Students' explanations should describe how the equation is used to relate the facts.

Try

Solve the problem.

① Marcie's older brother is away at college this year. She decided to save some of her weekly allowance to buy a camera for her computer so she can see her brother when she calls him. Her goal is to save $5 each week. One week she saved $8. What percent of her weekly goal did Marcie save that week?

Mark the Text

Read the Problem and Search for Information

Reread the problem to understand what you are asked to find.

Decide What to Do and Use Your Ideas

You know that Marcie's goal was to save __$5__ each week.
You also know that during one week she saved __$8__ instead of __$5__.
You are asked to find what __percent__ $8 is of her weekly goal.
You can use the strategy *Write an Equation* to solve the problem.

Ask Yourself: How can I identify the whole for the percent equation?

Step 1 Write a percent equation. Let p represent the percent that 8 is of 5.

percent × whole = part
$p \times 5 = 8$

Step 2 Solve the equation.
$p \times 5 = 8$
$5p = 8$
$p = 1.60$, or __100__ %

So Marcie saved __160__ % of her weekly goal.

Review Your Work

Reread the problem to be sure that you have not overlooked anything.

Determine Carlos says that Marcie saved 60% of her goal. What error might Carlos have made?

Sample Answer: The amount of increase is $3 and the percent of increase is 60%. Carlos forgot to add the 60% increase to the starting amount, which is 100%.

74 Unit 2 Using Proportional Reasoning

Guided Practice

Try

① Help students understand situations that involve percents greater than 100%.

Prompt students to look for the relationship between the goal and the amount saved.

- *How does the $8 Marcie saved compare with her weekly goal?*

Have students think about the percent Marcie saved.

- *Would you expect the percent Marcie saved to be less than or greater than 100%? Why?*

Ask students questions to make sure that they found what was asked for in the problem.

HOTS Determine Students should recognize that Carlos found the percent of increase, not the percent of Marcie's weekly goal.

Lesson 8 47

Scaffolded Practice
Apply

2 Encourage students to think carefully about the information given in the problem.
- Did the company spend more or less on dye this year? How did you decide?
- What equation can you write to find the change in spending?

HOTS Choose Responses should indicate that students multiplied to find the percent and then subtracted because the word *less* indicated subtraction.

3 Help students use clues from the problem to decide how to solve it.
- What words and numbers could you mark in the problem to help you decide what equations to use?

HOTS Plan Explanations should state that the original total is the whole and the difference between the two dollar amounts is the part. The percent is what students are to find.

4 Prompt students to think about the information given in relation to the question asked.
- Are there any numbers in the problem you do not need? How did you decide?
- How could you first write the equation in words to help you write the equation using numbers?

HOTS Compare Explanations should note that Problem 3 involves a decrease while this problem involves an increase, but both ask for the percent of change.

5 Guide students to consider the steps they will use to solve the problem.
- What will you do first? What equation can you write to find the amount of increase?

HOTS Explain Responses should state that the total amount is what the Reeds initially paid plus the cost of improvements.

48 Unit 2

Apply
Solve the problems.

2 Last year a company that makes action figures spent $2,060 for a special dye. This year, the dye costs 3% less than last year. How much did the company spend on the dye this year?

Whole × Percent reduced = Amount reduced
__2,060__ × __0.03__ = a

This year, the price of the dye is __$61.80__ less than it was last year.

Ask Yourself How can I find the amount by which the price was reduced?

Hint Three percent was subtracted from the price the company paid the year before.

Answer The company spent $1,998.20 on the dye this year.

Choose How did you decide which operations to use?
Sample Answer: I needed to multiply the percent by the whole and then subtract because there is a 3% decrease.

3 Students in Ms. Brown's math class used a computer program to play a stock market game. They started with an imaginary total of $18,000. At the end of two weeks, they had $16,470. By what percent was the starting total reduced?

The starting total was __$18,000__.
The reduced amount at the end of the two weeks was __$16,470__.
The starting total was reduced by __$1,530__.
Whole × Percent = Part
__18,000__ × p = __1,530__

Ask Yourself Do I know to what amount the price was reduced?

Hint In the percent equation, the whole (100%) is the original amount.

Answer The starting total was reduced by 8.5%.

Plan In the percent equation, how do you know which number is the whole, the part, and the percent?
Sample Answer: I used the original amount for the whole. I used the difference between the starting total and the reduced amount for the part. Then I could find the percent of change.

Lesson 8 Strategy Focus: Write an Equation 75

---

4 On his way to work on Monday, Mr. Parles put 14.5 gallons of gas in his car. He paid $2.50 per gallon. On Friday when he stopped at the same gas station, the price per gallon was $2.55. By what percent did the price per gallon increase?

Ask Yourself What is the whole in the percent equation?

The price of gas on Monday was __$2.50__ per gallon.
The price of gas on Friday was __$2.55__ per gallon.
The amount by which the price increased was __$0.05__.

Hint The amount of increase is the part in the percent equation.

Answer The price per gallon increased by 2%.

Compare How is this problem different from Problem 3? How are the two problems alike?
Sample Answer: Problem 3 is about a decrease; Problem 4 is about an increase. In both problems, I subtracted to find the amount of change before writing a percent equation.

5 The Reeds bought their house 5 years ago for $214,500. They have spent $28,000 on improvements. They want to sell the house for 115% of the total amount they paid, including improvements. What price should the Reeds ask for the house?

Ask Yourself Will the Reeds try to sell the house for more or for less than they paid?

The cost of the house was __$214,500__.
The amount spent on improvements is __$28,000__.
The total amount spent on the house is __$242,500__.

Hint Find the total amount the Reeds paid for the house with improvements.

Answer The Reeds should ask $278,875 for the house.

Explain Why do you need to add the cost of the house and the cost of improvements?
Sample Answer: I need to find 115% of the total amount that the Reeds spent on the house, including improvements.

76 Unit 2 Using Proportional Reasoning

Practice

Solve the problems. Show your work.

6. Last summer Taylor earned $264 by mowing lawns. He earned 120% of that amount during the winter by shoveling snow. How much money did Taylor earn by shoveling snow?

Some students will use the lesson strategy; however, other strategies may be used. Accept all reasonable work leading to the correct answer.

Answer Taylor earned $316.80 by shoveling snow.

Identify What equation can you use that will allow you to solve this problem in one step?

Sample Answer: I can use the equation 264 × 1.2 = a to solve the problem.

7. The owners of a pet adoption center spend about $3,500 each month to provide food for the animals they shelter. The owners spend 32% of this amount on food for cats. They estimate that the cost of cat food will increase by $30 a month next year. What is the percent of increase in the cost of cat food? Round your answer to the nearest tenth of one percent.

Some students will use the lesson strategy; however, other strategies may be used. Accept all reasonable work leading to the correct answer.

Answer The percent of increase in cat food is 2.7%.

Check How can you use estimation to check your work?

Sample Answer: I can estimate the cost of cat food as $\frac{1}{3}$ of $3,500, or about $1,200. An increase of $30 a month is then an increase of about $\frac{1}{40}$, or 2.5%.

Create Look at Problem 1. Change the savings goal and the amount saved. Write and solve a new problem with your numbers. See teacher notes.

Lesson 8 Strategy Focus: Write an Equation 77

Create

In this lesson, students modify Problem 1. They provide their own numbers for the savings goal and for the amount saved. Guide students to choose different whole numbers for the goal and for the amount saved.

Accept student responses that have made the changes described in the directions and that use an equation to find the correct answer.

Independent Practice
Practice

Students should be encouraged to choose any strategy to solve Problems 6 and 7, though many may prefer to use *Write an Equation*.

6. Some students may solve this problem by using a proportion.

Sample Work

whole × percent = part

$264 × 120% = amount Joe earned shoveling

$264 × 1.2 = $316.8

Identify Explanations should state that the percent equation could be applied as $264 × 120% = x or $264 × 1.2 = x.

7. Be sure students recognize this is a multi-step problem.

Sample Work

$3,500 × 32% = amount presently spent on food

$3,500 × 0.32 = 1,120

They presently spend $1,120.00 on food.

$1,120 × percent of increase = $30

percent of increase = $30 ÷ $1,120

≈ 0.027

Check Students' responses should describe how to use rounding and estimation to show their answer is reasonable.

Lesson 8 49

UNIT 2 Review

UNIT 2 Review

In this unit, you worked with three problem-solving strategies. You can often use more than one strategy to solve a problem. So if a strategy does not seem to be working, try a different one.

Problem-Solving Strategies
- ✓ Make a Table
- ✓ Write an Equation
- ✓ Work Backward

Check students' work throughout.
Students' choices of strategies may vary.
Solve each problem. Show your work. Record the strategy you use.

1. There were 455 students in the audience for the school play. That number of students was 125% of the number of adults in the audience. How many adults attended the play?

   Answer: __364 adults attended the play.__
   Strategy: __Possible Strategy: Write an Equation__

2. This year, the school soccer team made it to the state finals. Total attendance at all the home games was 38% greater than last year's attendance. If there were 912 more fans at the games this year, what was the attendance last year?

   Answer: __Attendance last year was 2,400 fans.__
   Strategy: __Possible Strategy: Write an Equation__

3. Paula makes necklaces to sell at craft shows. To make each necklace, she uses 30 small beads, 10 medium beads, and 5 large beads. On one day she used 315 beads in all making necklaces. How many large beads did Paula use that day?

   Answer: __Paula used 35 large beads.__
   Strategy: __Possible Strategy: Make a Table__

4. Since getting his new cell phone in April, Brett has increased the number of text messages that he sends by about 10% each month. In June, Brett sent 440 text messages. About how many text messages did Brett send in April?

   Answer: __Brett sent about 364 text messages in April.__
   Strategy: __Possible Strategy: Work Backward__

5. The school library had 2,645 books at the end of the school year. The number of new books bought during the year was 15% of the number of books the library had at the beginning of the year. The library did not lose or give away any books. How many books did the library have at the beginning of the year?

   Answer: __The library had 2,300 books at the beginning of the year.__
   Strategy: __Possible Strategy: Write an Equation__

   Explain how you solved this problem.

   Sample Answer: I let n be the number of books that the library had at the beginning of the year. I wrote the equation $2{,}645 = 1.15n$ because the number of books it had at the end of the year was 115% of the number of books it had at the beginning of the year. I solved the equation by dividing 2,645 by 1.15.

78 Unit 2 Using Proportional Reasoning
79

Support for Assessment

The problems on pages 78–81 reflect strategies and mathematics students used in the unit.

Although this unit focuses on three problem-solving strategies, students may use more than one strategy to solve the problems or use strategies different from the focus strategies. Provide additional support for those students who need it.

Make a Table For Problems 3, 7, and 10, help students set up a table if they choose to use one to organize their work. Make sure students think about how to organize the table so that they can answer the question asked in the problem.

Write an Equation For Problems 1, 2, 5, 6, and 9, suggest to students that they use the percent equation or a proportion. Prompt them to write equations correctly with questions such as, *Are you trying to find the percent, the whole, or the part?* For Problem 1, remind students that a percent greater than 100 means the part is greater than the whole. For Problem 9, help students realize that a decrease of 5% is the same as 95% of the original.

Work Backward For Problems 4 and 8, make sure students see that they are given an end value and are asked for a starting value. For Problem 4, make sure students realize they must find the number of messages for May before finding the number for April. For Problem 8, help students sequence the steps with questions such as, *Should you subtract the bonus before or after you take into account the 2.5% raise?*

You may wish to use the *Review* to assess student progress or as a comprehensive review of the unit.

50

Promoting 21st Century Skills

Write About It
Communication
When students are asked to describe how they chose a strategy, they have an opportunity to reflect on how the information in the problem relates to what they need to find. Students should clearly describe how they chose a problem-solving strategy.

Team Project: Prizes for Field Day
Collaboration: 3–4 students

Emphasize that in a group project, the ideas of each member are important and should be considered when making a decision. Guide students to begin by discussing how the 10% discount affects the amount they can spend. Suggest that they individually decide the number of each type of prize to order, as well as a format for the order form. Then they should come to a consensus after sharing their ideas.

Ask questions that help students summarize their thinking. *How did you take into account the 10% discount? How did you decide what information should be included on the order form?*

Extend the Learning
Media Literacy
If you have Internet access, navigate to sites where students can reconsider the problem using the actual cost of trophies and ribbons.

🔍 Search **trophies and ribbons**

Solve each problem. Show your work. Record the strategy you use.

6. The 60 members of the school band marched in the Town Day Parade. If 12% of all the people who were in the parade marched in the school band, how many people were in the parade?

   Answer _500 people were in the parade._
   Strategy _Possible Strategy: Write an Equation_

7. Pete and Mara are both training for a long bicycle race. On Day 1, Pete cycles 20 miles and Mara cycles 18 miles. Each day, Pete increases his distance by 10% and Mara increases her distance by 15%. On which day of training will Mara cycle a greater distance than Pete?

   Answer _Mara will cycle a greater distance than Pete on Day 4._
   Strategy _Possible Strategy: Make a Table_

8. Monty received a raise of 2.5%. His latest paycheck was for $761.25, which included a one-time bonus of $300. Before his raise, how much did Monty earn each week?

   Answer _Monty earned $450.00 before his raise._
   Strategy _Possible Strategy: Work Backward_

   Explain how you could solve this problem using a different strategy.

   Sample Answer: I could write an equation. If x is the amount of weekly pay, in dollars, before the raise, then $1.025x + 300 = 761.25$.

9. The number of students enrolled in the middle school this year is 5% lower than last year's enrollment. The enrollment this year is 855 students. How many students were enrolled last year?

   Answer _There were 900 students enrolled last year._
   Strategy _Possible Strategy: Write an Equation_

10. Julian earned $172.50 last weekend. When he works more than 8 hours in one day, Julian earns 1.5 times his regular hourly rate for the overtime hours. He worked 5 hours on Friday, 10 on Saturday, and 7 on Sunday. What is his regular hourly rate of pay?

    Answer _His regular hourly rate is $7.50._
    Strategy _Possible Strategy: Make a Table_

Write About It
Look back at Problem 3. Describe how you used the information in the problem to choose a strategy for solving the problem.

Sample Answer: I saw there was a ratio of different-sized beads to the total number of beads used, so I used the strategy Make a Table to find how many large beads Paula used in all.

Team Project: Prizes for Field Day

Your committee has been asked to order prizes for the school Field Day. There are 10 events for which prizes will be awarded. You have a budget of $50. There is a 10% discount on orders of $40 or more.

Your goal is to spend as much of your budget as possible.

Prices for Prizes	
Large trophy	$8.75
Small trophy	$6.50
Medal	$1.50
Ribbon	$.75

Plan You need first- and second-place prizes for ten events.
Decide As a group, choose which prizes to order and how many of each.
Create Make an order form that lists each prize, the number and cost of each, and the total cost, including any discount.
Present As a group, share your plan and your order with the class. Include how much money is left from the $50 budget.

Unit 2 51

UNIT 3 Problem Solving Using Algebra

CCSS 8.NS Number Sense | CCSS 8.G Geometry

Unit Overview

Lesson	Problem-Solving Strategy	Math Focus
9	Make a Graph	Graphing on the Coordinate Plane
10	Make a Table	Linear Functions
11	Guess, Check, and Revise	Pythagorean Theorem
12	Draw a Diagram	Transformations

Promoting Critical Thinking

Higher order thinking questions occur throughout the unit and are identified by this icon: HOTS. These questions progress through the cognitive processes of remembering, understanding, applying, analyzing, evaluating, and creating to engage students at all levels of critical thinking.

UNIT 3 Problem Solving Using Algebra

Unit Theme: Working in Our World

Planning a city's layout, constructing buildings, and creating works of art all require a lot of thought and effort. In this unit, you will see how people use math in all sorts of ways when working in our world.

Math to Know

In this unit, you will apply these math skills:
- Graph points and lines on a plane
- Use the Pythagorean Theorem
- Use transformations in a plane

Problem-Solving Strategies
- Make a Graph
- Make a Table
- Guess, Check, and Revise
- Draw a Diagram

Link to the Theme

Write another paragraph about the guided walking tour that Isaac's family wants to take. Include some of the words and numbers from the table.

Isaac's family is on vacation. They want to take a walking tour of the city. The sign at the information booth lists the tour information.

Tour	Distance (miles)	Price (per person)
History Trail	$2\frac{1}{4}$	$5
Arts Trail	3	$7
Sports Trail	$4\frac{1}{4}$	$10

Students' paragraphs will vary, but should include some words and numbers from the table.

Use Math Language

Review Vocabulary

The list below shows vocabulary terms in this unit. Knowing the meaning of these terms will help you understand the problems.

hypotenuse   Pythagorean Theorem   rotation   transformation
leg   reflection   slope   translation

Vocabulary Activity — Multiple-Meaning Words

Some math terms are also used in everyday English. Use words from the list above to complete each sentence.

1. In a right triangle, one __leg__ forms a right angle with another __leg__.
2. Sue had to stop running because she got a cramp in her __leg__.
3. A horizontal line has a __slope__ of zero.
4. Zach could not wait to ski down the snow-covered __slope__.

Graphic Organizer — Word Charts

Complete the graphic organizer.
- Cross out the word that does not belong.
- Replace it with a word from the vocabulary list that does belong.
- Write a definition for each word.

Sample Answers:

reflection	~~hypotenuse~~ translation
a movement of a figure by flipping it across a line	a movement of a figure by sliding it
rotation	transformation
a movement of a figure by turning it around a fixed point	a change of a figure in its position, shape, size, and/or orientation

---

Link to the Theme   Working in Our World

Ask students to read the direction line and story starter. If students are having trouble getting started, ask questions such as, *How much longer is the Arts Trail than the History Trail? How many times more expensive is the Sports Trail tour than the History Trail tour?*

Unit 3 Differentiated Instruction

Extra Support

Some students may need to review the use of formulas before they apply them in a problem-solving situation.

Area and Perimeter Have students find the perimeter and area of a rectangle on a grid by counting. Then review the formulas for finding these measures. Repeat the process with triangles. For the area, point out that a right triangle is half of a rectangle, so the area is $\frac{1}{2} \times$ length $\times$ width, where the length and width are the legs of the right triangle.

Pythagorean Theorem Sketch and label a right triangle with legs 12 cm and 16 cm. Discuss how the parts of the triangle are used in the Pythagorean Theorem. Remind students to use the order of operations as they find the length of the hypotenuse.

Challenge Early Finishers

Students who finish early may enjoy problems with more than one possible answer. Show students a diagram such as the one below.

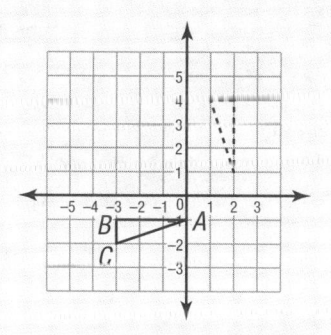

Have them describe a set of transformations that moves the dashed triangle into the same position as the solid one. Once students are sure their transformations work, ask them to find other sets of transformations that give the same results. The new solutions should change more than just the order of the transformations.

English Language Learners

Listening and Speaking

Horizontal, Vertical, Diagonal Draw a horizontal line on the board, point to it, and say, *This is a horizontal line*. Have students repeat after you. Then do the same with a vertical line and a diagonal line. To provide students with additional practice using these terms, give pairs of students graph paper and have one student say, for example, *Draw a vertical line*. The other student follows the direction by drawing a vertical line. The first student continues giving directions about drawing lines that include the words *horizontal*, *vertical*, and *diagonal* until the second student has drawn a design on the graph paper. Have students switch roles and repeat.

Build Background

Real Objects To give meaning to the problems, show pictures of the architectural features described in the unit. For example, find in books or on the Internet pictures of a Pratt truss bridge, a gable end of a pitched roof, and a wall brace. Have students describe these objects in their own words, and discuss how the pictures relate to the diagrams provided with the problems.

Reading Comprehension

Multiple-Meaning Words Write the following sentences from the unit on the board: *What is the least amount of time, to the nearest second, in which 7 cars can pass through the intersection?* and *If they travel at the same rate, how long will it take them to drive from their first destination to their second destination?* Circle the word *second* in each sentence, and remind students that some words have more than one meaning. Discuss that in the first sentence, *second* refers to a unit of time, and in the second sentence, *second* is used as a sequence word. Then help students find the two different meanings of the word *left* as it is used in the unit. Guide students to recognize that in some sentences, *left* is used as a direction word, and in other sentences, *left* means something is remaining.

Lesson 9: Strategy Focus — Make a Graph

Lesson Overview

Lesson Materials: calculator

Skills to Know	Outcome	eResources — www.optionspublishing.com
• Graph points on a coordinate grid • Read information from a graph	Students will recognize that graphs are an efficient way to predict a value based on a known relationship between data.	• Interactive Whiteboard Transparency 9 • Homework, Unit 3 Lesson 9 • First Quadrant Grids • Problem-Solving Checklist, also available in the student worktext, page 7

Modeled Instruction

Learn

Probe students' understanding of the problem's context by asking questions similar to the following.

- Why would a city planner want to know how quickly cars can pass through an intersection?
- What does *maximum* mean?

As students read the problem again, ask questions to help them focus on the details needed to solve the problem.

- If 3 cars can pass through the intersection in 2 seconds, how many cars could pass through the intersection in 4 seconds?

Reread You may wish to use this Think Aloud to demonstrate to students how to present the facts in the problem in a way that helps them answer the question.

This problem is about a city planner studying the numbers of cars that go through an intersection in a certain amount of time. It says that in 2 seconds, a maximum of 3 cars can pass through the intersection. I need to find the least amount of time it would take 7 cars to go through.

I could use a pattern: 3 cars in 2 seconds; 6 cars in 4 seconds; 9 cars in 6 seconds…No, that will not work because when I count by threes I miss 7. But now I know that the answer is somewhere between 4 and 6 seconds. I need to find a way that shows the in-between numbers. Hey, I've got an idea! I can show the pattern on a graph that compares time and cars. Then I can look to see how long it would take 7 cars to go through the intersection.

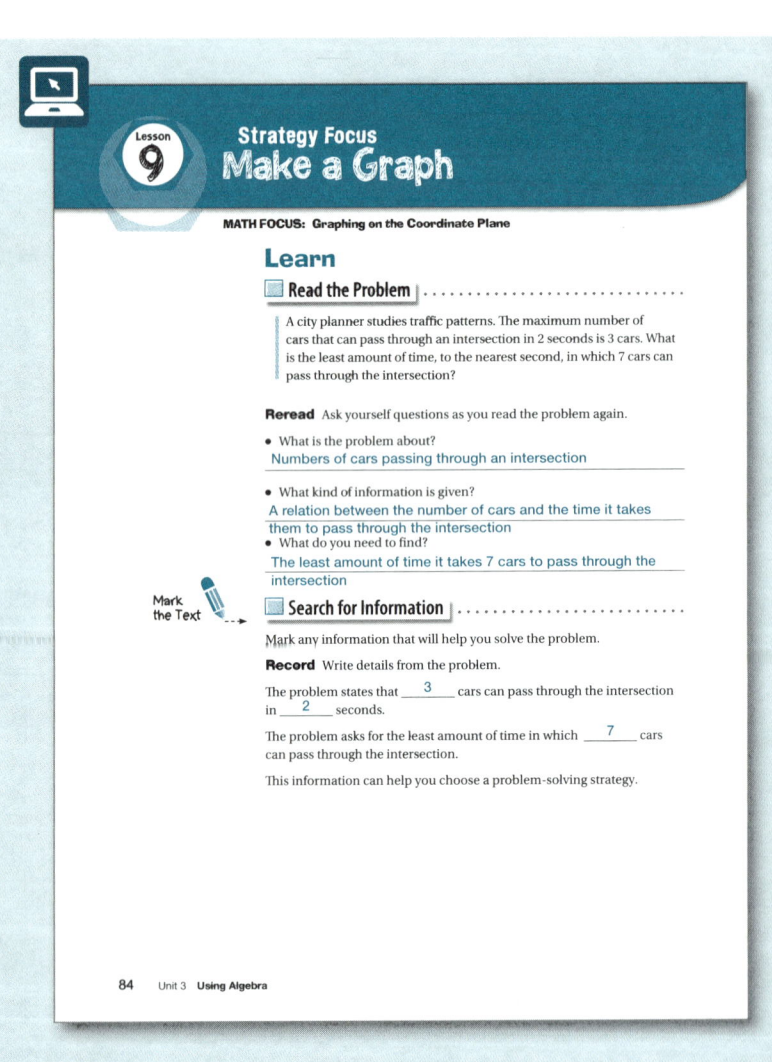

Lesson 9: Strategy Focus — Make a Graph

MATH FOCUS: Graphing on the Coordinate Plane

Learn

Read the Problem

A city planner studies traffic patterns. The maximum number of cars that can pass through an intersection in 2 seconds is 3 cars. What is the least amount of time, to the nearest second, in which 7 cars can pass through the intersection?

Reread Ask yourself questions as you read the problem again.

- What is the problem about?
  Numbers of cars passing through an intersection
- What kind of information is given?
  A relation between the number of cars and the time it takes them to pass through the intersection
- What do you need to find?
  The least amount of time it takes 7 cars to pass through the intersection

Mark the Text

Search for Information

Mark any information that will help you solve the problem.

Record Write details from the problem.

The problem states that __3__ cars can pass through the intersection in __2__ seconds.

The problem asks for the least amount of time in which __7__ cars can pass through the intersection.

This information can help you choose a problem-solving strategy.

84   Unit 3 Using Algebra

---

54   Unit 3

Student Page 85

Decide What to Do

You know the relationship between the number of cars passing through the intersection and the number of seconds that it takes.

Ask How can I find the least amount of time in which 7 cars can pass through the intersection?

- I can use the strategy *Make a Graph* to graph a line showing the relationship between time passed and the number of cars.
- The line will show the greatest number of cars that can pass through the intersection over time.

Use Your Ideas

Step 1 Make a table. Start with the data points you know and count up by 2 seconds and 3 cars to complete the table.

Time (seconds)	0	2	4	6
Number of Cars	0	3	6	9

After 0 seconds, 0 cars have passed through the intersection.

Step 2 Set up a graph. Label the vertical axis *Number of Cars* and the horizontal axis *Time (seconds)*.

Step 3 Plot the points on the graph. Draw a line through the points to the edge of the graph.

Step 4 Find the point on the line that represents 7 cars. That point is at about ___5___ seconds.

To the nearest second, 7 cars can pass through the intersection in about ___5___ seconds.

Review Your Work

Check that you have plotted the points correctly.

Describe How can you use your graph to find how long it takes 17 cars to pass through the intersection?
Sample Answer: I could extend the graph and find the point on the line where the number of cars is 17. Then I could draw a vertical line from that point to the x-axis to see how many seconds it would take.

Student Page 86

Try

Solve the problem.

1. A public works department of a city employs 30 people. The population of the city is 10,000. The department has found that it needs 1 additional employee for every additional 500 people who live in the city. What will be the population of the city when 100 employees are needed? Give your answer to the nearest 5,000 people.

Read the Problem and Search for Information

Mark any information that will help you solve this problem.

Decide What to Do and Use Your Ideas

You can use the strategy *Make a Graph* to solve the problem.

Step 1 Make a table to organize the data.

Population (thousands)	10	20	30
Employees	30	50	70

Ask Yourself If 1 additional employee is needed for each additional 500 people, how many additional employees are needed for each additional 5,000 people?

Step 2 Plot the points. Draw a line through the points.

Step 3 Find where the line shows 100 employees. The population that corresponds to that point is about ___45,000___.

So the city will need 100 employees when it has a population of about ___45,000___ people.

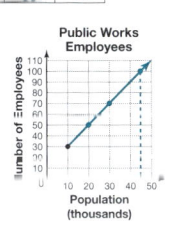

Review Your Work

Make sure you have rounded your answer to the nearest 5,000.

Recognize Why is making a graph a good strategy to solve this problem?
Sample Answer: Making a graph works because the solution can be read from the graph.

Modeled Instruction (continued)

Help students connect the facts they know to a strategy they can use to solve the problem.

- *How can you use a graph to display the relationship between number of cars and time?*

Ask questions that guide students to consider each step in the solution process.

- *Why do you add 3 to the number of cars every time that the time increases 2 seconds?*
- *How can you use the graph to find the number of seconds that corresponds to 7 cars?*

Emphasize to students the importance of graphing the points and drawing the line accurately.

HOTS Describe Students' explanations should demonstrate an understanding that they can extend the line to extend the pattern and find the answer.

Guided Practice

Try

1. Emphasize to students that the problem relates the number of residents to the number of employees.

   Make sure students can distinguish between employees and residents.

   - *Are you trying to find the number of employees or the number of residents of the city?*

   Help students complete the table to make the graph.

   - *What does the point (10, 30) represent?*

   Make sure students rounded their answers correctly.

   - *What does it mean to round your answer to the nearest 5,000?*

   HOTS Recognize Responses should note that a graph clearly shows the relationship between the variables.

Lesson 9   55

Scaffolded Practice
Apply

2 Ask questions to help students understand the situation described in the problem.
- *How are the words* minimum *and the least number related?*
- *If you get a fraction or a decimal for the number of hydrants, should you round up or round down?*

HOTS Explain Students' responses should point out that labeling the axes helps them make sure the information is plotted correctly and that the graph can be easily read.

3 Make sure students understand how the table is set up.
- *Why does the amount of money decrease as the number of trees increases?*
- *How else could you set up a table to solve the problem?*

HOTS Determine Explanations should note that Marta did not use all the necessary details in the problem. She did not leave any money for a separate planting.

4 Help students interpret the meaning of a non-whole number answer in the context of the problem.
- *Should you round up or round down if your answer is not a whole number? Why?*

HOTS Identify Explanations should indicate that the table helps students organize the coordinates for the points on the graph.

5 Make sure students understand how to choose the intervals on the scales for the graph.
- *Would you want to use increments of 1 on the graph? Why or why not?*

HOTS Discuss Responses should show that students can use the graph in other ways by finding a money amount instead of a time.

Apply
Solve the problems.

2 In Brownsville, a minimum of 2 fire hydrants are required for every 3 blocks. What is the least number of hydrants needed for 8 blocks?

Number of Blocks	Number of Hydrants
3	2
6	4
9	6

Fire Hydrants per Block graph

Ask Yourself Will my answer be a mixed number or a whole number?

Hint The minimum number of hydrants will always fall on or above the line, never below it.

Answer At least 6 hydrants are needed for 8 blocks.

Explain Why is it important to clearly label the axes of a graph?
Sample Answer: Labeling the axes makes it clear what elements you are relating.

3 The parks department has $4,200 to spend on trees. It can buy 10 trees for $400. How many trees can the department buy if it wants to leave $1,000 for a separate planting later this year?

Trees Purchased	Amount Left ($)
0	4,200
10	3,800
20	3,400
30	3,000

Trees for Parks Department graph

Hint Find coordinates for three more points on the graph to help you solve the problem.

Ask Yourself What does the line in the graph represent?

Answer The parks department can buy 80 trees and still have $1,000 left.

Determine Marta says the answer is 105 trees. What mistake did she probably make?
Sample Answer: Marta probably forgot to leave $1,000 for a separate planting.

Lesson 9 Strategy Focus: Make a Graph 87

4 A town has found that at least 3 apartment buildings are required to support each business in a neighborhood. What is the greatest number of businesses that 8 apartment buildings can support?

Number of Apartment Buildings	Number of Businesses
0	0
3	1
6	2
9	3

Businesses per Apartment Building graph

Ask Yourself How can I use what I know to complete the table?

Hint The points below the line of the graph show numbers of businesses and apartment buildings that relate in the recommended way.

Ask Yourself What point on the graph do I need to look at for the solution?

Answer An area with 8 apartment buildings can only support 2 businesses.

Identify How does making a table help you make a graph?
Sample Answer: The table helps me find the points that I will plot to graph a line.

5 A town has an annual budget of $455,000 for snow plowing. For every 1,000 hours the crew spends plowing snow, it costs $25,000. How many hours can the town afford before it runs out of money? Round to the nearest thousand hours.

Time Spent (thousands of hours)	Money Left (thousands of dollars)
0	455
2	405
4	355
6	305

Snow Removal Budget graph

Hint Be sure to plot your points correctly.

Answer The town can afford about 18,000 hours of snow removal.

Discuss How can you use the graph to find out how much money will be left in the budget after 12,000 hours of snow removal?
Sample Answer: I can find the point on the line that represents 12,000 hours and see what amount of money is left.

Practice

Solve the problems. Show your work.

6 A city plan includes a minimum number of parks per square mile. For each 4 square miles, there should be at least 3 parks. How many parks are needed for 12 square miles?

Some students will use the lesson strategy; however, other strategies may be used. Accept all reasonable work leading to the correct answer.

Answer For 12 square miles, 9 parks are needed.

Conclude How does a graph help you solve a problem like this?

Sample Answer: I can plot given information to graph a line. Then I can extend the line to find information that is not given.

7 A city with a population of 40,000 people has 5 city council members. For every additional 20,000 people who live in the city, the council must add 1 new member. How many people will be on the council if the population is 280,000?

Some students will use the lesson strategy; however, other strategies may be used. Accept all reasonable work leading to the correct answer.

Answer The city council will have 17 members if the population is 280,000.

Compare How is this problem like Problem 1?

Sample Answer: Both problems have a rate that depends on the population to determine how many people are needed for a certain job.

Create Two cars are allowed over a small bridge every 5 seconds. Use this information to write a new problem that can be solved by using the strategy *Make a Graph*. See teacher notes.

Lesson 9 Strategy Focus: Make a Graph 89

Create In this lesson, students create a problem using the information provided. If students are struggling, suggest they look back at the first problem in the lesson and write a similar problem. Encourage them to choose a number of seconds that is not a multiple of 5 so they will have to interpret a non-whole number.

Accept student responses that ask a question based on the given information and that provide a correct solution that uses a graph.

Independent Practice
Practice

Students should be encouraged to choose any strategy to solve Problems 6 and 7, though many may prefer to use *Make a Graph*.

6 Some students may choose to solve this problem directly with an equation.

Sample Work

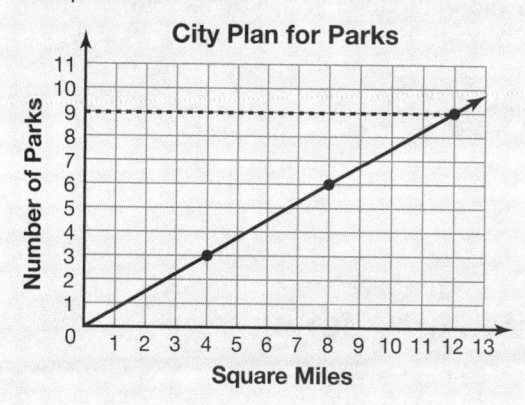

City Plan for Parks

(graph: Number of Parks vs. Square Miles, points at (4,3), (8,6), (12,9))

Conclude Students' responses should show that the graph allows them to find specific points on a graph.

7 Make sure students understand the relationship between the two variables.

Sample Work

City Council Membership

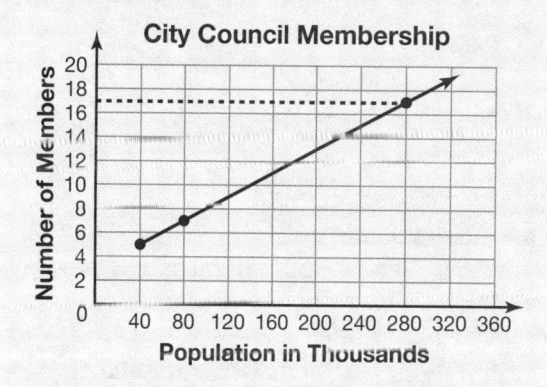

(graph: Number of Members vs. Population in Thousands, points at (40,5), (80,7), (280,17))

Compare Students' responses should indicate that for Problems 1 and 7, the contexts and the way the problems are solved are similar.

Lesson 9   57

Lesson 10

Strategy Focus: Make a Table

Lesson Overview

Lesson Materials: calculator

eResources: www.optionspublishing.com

Skills to Know	Outcome	Math Vocabulary	eResources
• Read information from a graph • Find the slope of a line • Find and apply a pattern rule	Students will recognize that tables are an efficient way to solve problems involving a constant rate of change.	slope	• Interactive Whiteboard Transparency 10 • Homework, Unit 3 Lesson 10 • Know-Find Table • Problem-Solving Checklist, also available in the student worktext, page 7

Modeled Instruction

Learn

To be sure students understand the context of the problem, ask questions such as the ones below.

- *What are the two parts that make up the shuttle bus fee?*
- *Which part of the fee is set? Which part varies?*

As students read the problem again, guide them to identify the words and numbers needed to solve the problem.

- *What does each point on the graph represent in the context of the problem?*
- *What two questions do you need to answer?*

Use a Graphic Organizer You may wish to use this Think Aloud to demonstrate how to organize information obtained from a problem and its graph.

In this problem, I need to find how much a shuttle bus charges per mile in addition to the boarding fee. I need to get information from both the problem and a graph. I think I will use a Know-Find Table to keep track of the information.

In the Know column, I will write The boarding fee is $5.00. *I will also add information from the graph:* Traveling 0 miles costs $5 *and* Traveling 4 miles costs $8.00. *In the Find column, I will write the questions given in the problem:* What is the set rate for each additional mile traveled? *and* How much would it cost to travel 16 miles?

I think I need to use another table to help me solve this problem.

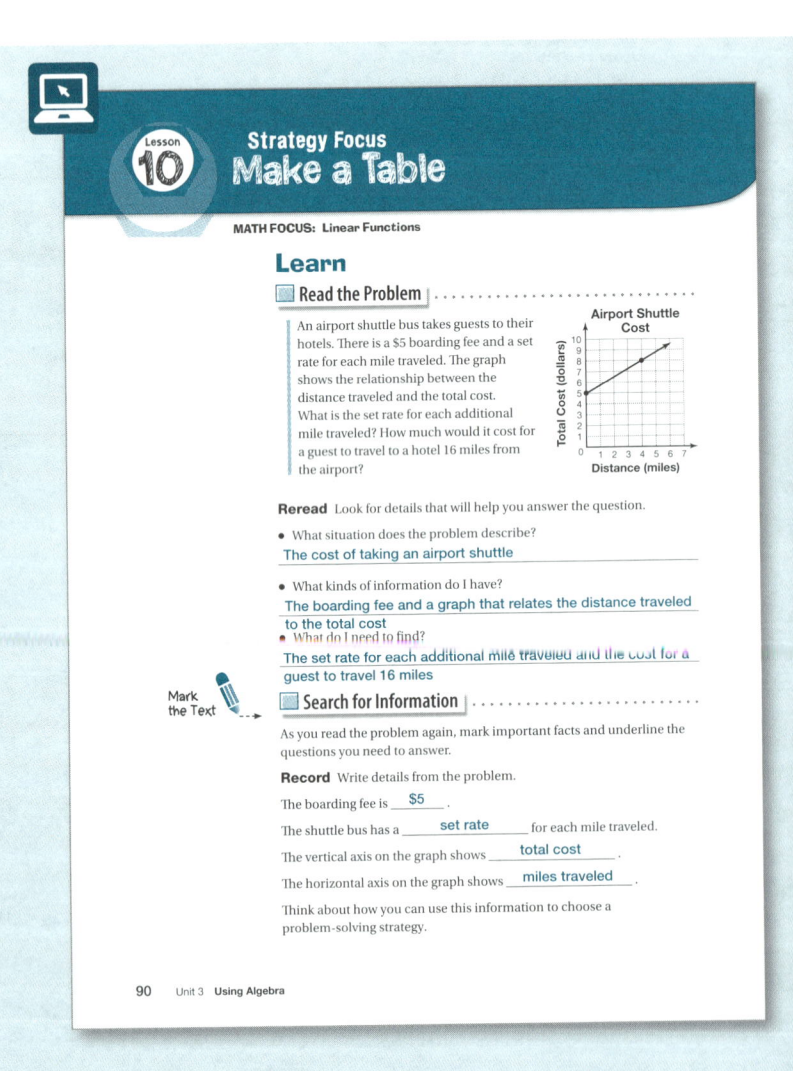

58 Unit 3

Decide What to Do

You know the cost of boarding the shuttle bus. You know the graph relates the distance traveled to the total cost.

Ask How can I find the set rate for each additional mile and the total cost to travel 16 miles?

- I can use the graph to find the *slope*, which will tell me the set rate for each additional mile.
- Then, I can use the strategy *Make a Table*. I can extend the pattern in the table to find the cost of traveling 16 miles.

Use Your Ideas

Step 1 Find the set rate for each additional mile traveled. Use the slope formula.

slope → $\frac{\text{change in cost}}{\text{change in distance}} = \frac{8-5}{4-0}$

$= \frac{3}{4}$

So the set rate for each additional mile traveled is $\frac{3}{4}$ of a dollar, or $ 0.75.

Slope tells the rate of change.

Step 2 Make a table to show the costs for different trips. Write the distance in the top row and the total cost in the bottom row.

For every 4 miles the distance increases, the total cost increases by $3.

Distance (miles)	0	4	8	12	16
Total Cost (dollars)	5	8	11	14	17

Step 3 Extend the pattern in the table to find the total cost for a 16-mile trip.

So the cost of a 16-mile trip is $17.

Review Your Work

Make sure you used the correct pattern when completing the table.

Describe How did making the table help you solve the problem?

Sample Answer: The table helped me extend the pattern to get information not shown on the graph.

Modeled Instruction (continued)

Help students recognize how they can determine a problem-solving strategy to use.

- *What two values are compared in the graph?*
- *How can you record the coordinates of the points so you can look for and use a pattern?*

Ask questions that guide students to consider each step in the solution process.

- *Why do you use the change in cost in the numerator of the slope formula?*
- *What pattern helps you complete the table?*

Emphasize the importance of using the correct pattern when completing the table.

HOTS Describe Students' explanations should note that the table allowed them to find and extend the pattern and organize the values they calculated.

Try

Solve the problem.

1. Monica's family is going on a road trip. The graph shows the distance they have left to drive to their first destination after various numbers of hours. Their second destination is 350 miles from their first destination. If they travel at the same rate, how long will it take them to drive from their first destination to their second destination?

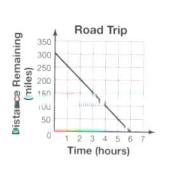

Mark the Text

Read the Problem and Search for Information

Identify what you know about the distances the family traveled.

Decide What to Do and Use Your Ideas

Make a table to find a pattern.

Ask Yourself: If I know how far the family has left to travel at various times, how can I find how far the family traveled each hour?

Step 1 Make a table to show the total distance traveled after different amounts of time. Use the distance remaining each hour to find the distance traveled each hour. Then continue the pattern.

Distance traveled in 1 hour: 300 − 250 = 50

Distance traveled in 2 hours: 300 − 200 = 100

Time (hours)	1	2	3	4	5	6	7
Distance (miles)	50	100	150	200	250	300	350

Step 2 Extend the pattern until you find the time it takes to drive 350 miles.

So it will take the family 7 hours to get from their first destination to their second destination.

Review Your Work

Does the slope of the graph match the rate shown in the table?

Relate What phrase in the problem helps you know you can find and extend a pattern? Explain.

Sample Answer: The phrase *at the same rate* means the number of miles traveled each hour is constant. That means there is a pattern in the number of miles traveled each hour.

Guided Practice

Try

1. Ask questions that help students understand the problem.

   Have students consider the information from the graph in the context of the problem.

   - *Why is the line on the graph going downhill as the number of hours the family travels increases?*
   - *How are the distances of the two parts of their trip different?*

   Help students understand the table in the context of the problem.

   - *Why do you subtract to find the miles driven?*
   - *What is the family's rate of travel in miles per hour?*

   Have students check their answers by calculating the slope of the graph and using it to complete the table.

   HOTS Relate Responses should indicate that the phrase *at the same rate* implies that the speed is constant.

Lesson 10

Scaffolded Practice
Apply

② Guide students to use the concept of slope to solve the problem.
- How can you find and use the slope of the graph to complete the table?
- What unit rate does the slope represent?

HOTS Identify Explanations should show that students realize that the speed of the car is not needed to solve the problem.

③ Prompt students to explain how they will decide what numbers to use in the table.
- How can you use information in the graph to fill in the first three columns in the table?
- How can you use the numbers in the first three columns to complete the table?

HOTS Summarize Students' responses should demonstrate an understanding that they can extend the graph to find the solution.

④ After completing the table, ask students to relate the information they recorded to the question that is asked.
- How can you use the numbers in the table to determine your answer?

HOTS Explain Students' responses should note that the table helps them solve the problem because it allows them to organize values of time and distance as they extend the pattern.

⑤ Help students see how the data in the graph are related to the data in the table.
- How do you use the information from the graph to find the number of gallons remaining?

HOTS Determine Students' responses should show an understanding of how the information given in the problem and the graph can be used to answer other questions.

60   Unit 3

Apply

Solve the problems.

② The graph shows the amount of gas used when a car travels at a constant rate of 50 miles per hour. How many gallons will the car use in 10 hours if it is traveling at a constant rate of 50 miles per hour?

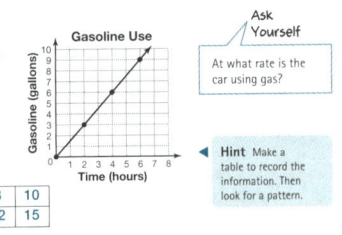

Ask Yourself At what rate is the car using gas?

Gasoline Use graph

Hint Make a table to record the information. Then look for a pattern.

Time (hours)	0	2	4	6	8	10
Gas (gallons)	0	3	6	9	12	15

Answer The car will use 15 gallons in 10 hours.

Identify What information is given that is not needed to solve the problem?

Sample Answer: You do not need to know that the car travels at 50 miles per hour.

③ A grocery service makes deliveries at the rate shown on the graph. At this rate, how many minutes will it take to make 21 deliveries?

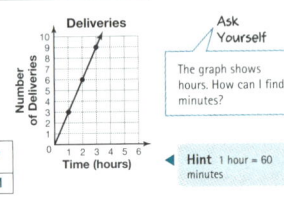

Ask Yourself The graph shows hours. How can I find minutes?

Deliveries graph

Hint 1 hour = 60 minutes

Time (hours)	1	2	3	4	5	6	7
Deliveries	3	6	9	12	15	18	21

Answer It will take 420 minutes to make 21 deliveries.

Summarize Suppose you transferred the graph onto a larger sheet of paper. What is another way you might solve the problem?

Sample Answer: I could extend the line and read the corresponding time for 21 deliveries.

Lesson 10 Strategy Focus: Make a Table   93

---

④ Sonja has 10 minutes to deliver a document to a building that is 4.5 kilometers away. The graph shows the distance she has traveled during the first 4 minutes on her way to the building. If she bikes at the same speed as in the first 4 minutes, will she make it in time? Explain.

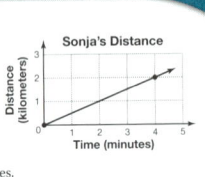

Ask Yourself What unit could you use to describe her speed?

Sonja's Distance graph

Hint Use the graph to find the slope. Then use the slope to make the table.

Time (minutes)	2	4	6	8	10
Distance (kilometers)	1	2	3	4	5

Answer Yes, Sonja will reach the building in 9 minutes.

Explain How does the table help you solve the problem?

Sample Answer: Since 4.5 kilometers is halfway between 4 and 5 kilometers, the time is halfway between 8 and 10 minutes.

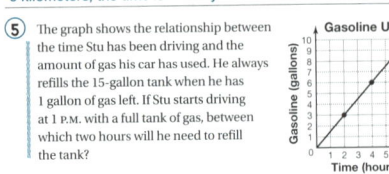

⑤ The graph shows the relationship between the time Stu has been driving and the amount of gas his car has used. He always refills the 15-gallon tank when he has 1 gallon of gas left. If Stu starts driving at 1 P.M. with a full tank of gas, between which two hours will he need to refill the tank?

Hint You are looking for two specific hours of the day, not just the time elapsed.

Ask Yourself How does the graph help me figure out what values to put into the table?

Gasoline Use graph

Time of Day (P.M.)	1	3	5	7	9	11
Gallons Remaining	15	12	9	6	3	0

Answer Stu will need to refill the tank between 10 P.M. and 11 P.M.

Determine What is another question you could ask from the information in the problem?

Sample Answer: At what rate does the car use gas?

94   Unit 3 Using Algebra

Practice

Solve the problems. Show your work.

6 The driver of a horse-drawn carriage charges an initial fee and a set rate for each block traveled. The graph shows the relationship between the distance traveled and the total cost. If Marcia travels 7 blocks from her starting point, how much will her trip cost?

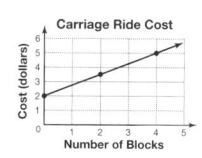

Carriage Ride Cost

Some students will use the lesson strategy; however, other strategies may be used. Accept all reasonable work leading to the correct answer.

Answer Marcia's trip will cost $7.25.

Analyze Mike says the cost of the trip will be $5.25. What mistake did Mike probably make?

Sample Answer: Mike did not include the initial fee of $2.00 in the cost of the trip.

7 Randy drives at a constant speed. The graph shows the distance that he has traveled during the first 60 minutes. How long will it take Randy to drive 270 miles?

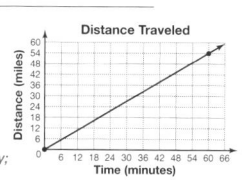

Distance Traveled

Some students will use the lesson strategy; however, other strategies may be used. Accept all reasonable work leading to the correct answer.

Answer It will take Randy 5 hours to drive 270 miles.

Outline Describe another method you could use to solve the problem.

Sample Answer: I could use the graph to find the average speed of 54 miles per hour. Then I could divide 270 miles by the average speed to find it will take 5 hours.

Create Look back at the problems in the lesson. Choose one, and change the slope of the graph. Write and solve your new problem. Be sure it can be solved by using the strategy *Make a Table*.

See teacher notes.

Lesson 10 Strategy Focus: Make a Table   95

Create

In this lesson, students modify an existing problem. They need to choose a problem from the lesson and redraw the line using a different slope. If students are struggling, suggest they keep the *y*-intercept the same and choose another point on the grid for the line to pass through. Make sure they draw a line with constant slope.

Accept student responses that provide a problem that is similar to one in the lesson, but with a different slope. The solution should include a table based on information from the graph.

Independent Practice
Practice

Students should be encouraged to choose any strategy to solve Problems 6 and 7, though many may prefer to use *Make a Table*.

6 Some students may choose to extend the graph rather than using a table to solve the problem.

Sample Work

Blocks	0	2	4	6	8
Cost	$2.00	$3.50	$5.00	$6.50	$8.00

$$\frac{\$6.50 + \$8.00}{2} = \$7.25$$

Analyze Responses should show that students can recognize common errors. They should note that Mike did not include the initial fee.

7 Encourage students to change the units of time to hours so the numbers are easier to work with.

Sample Work

Hours	0	1	2	3	4	5
Miles	0	54	108	162	216	270

Outline Students' explanations should note that they can use the rate to find the answer.

Lesson 10   61

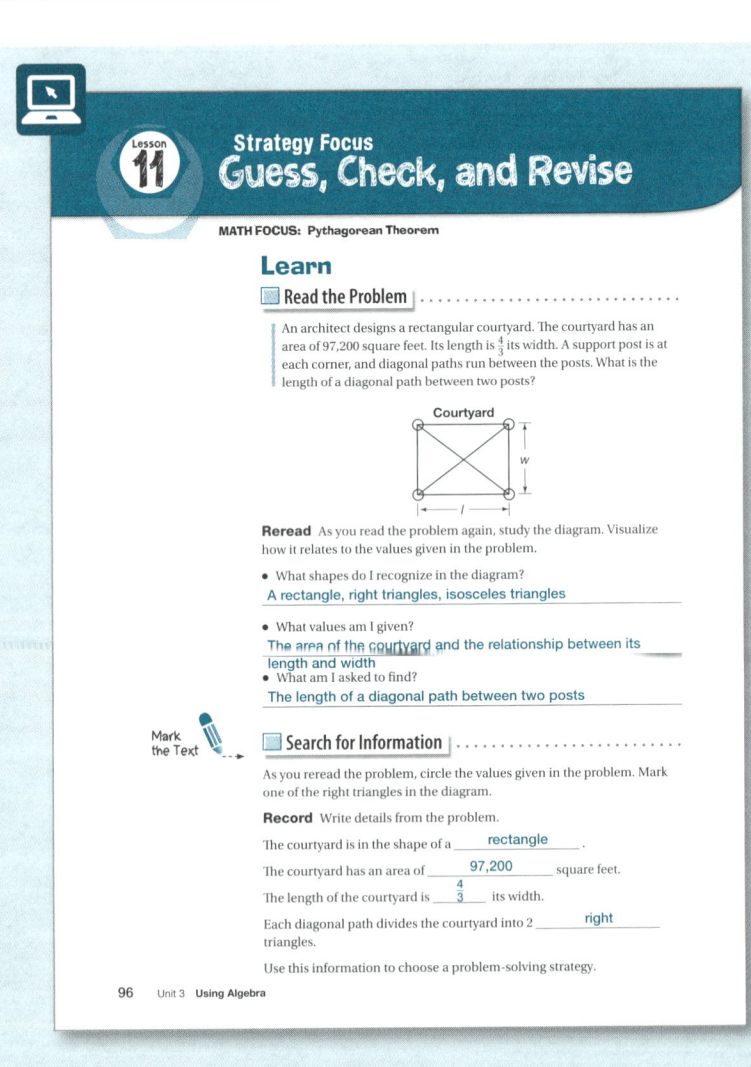

Lesson 11

Strategy Focus
Guess, Check, and Revise

Lesson Overview

Lesson Materials: calculator

Skills to Know	Outcome	Math Vocabulary	eResources www.optionspublishing.com
• Use area and perimeter formulas • Use the Pythagorean Theorem	Students will recognize that guessing, checking, and revising is an efficient way to solve problems involving stated relationships among quantities.	hypotenuse, leg, Pythagorean Theorem	• Interactive Whiteboard Transparency 11 • Homework, Unit 3 Lesson 11 • Problem-Solving Checklist, also available in the student worktext, page 7

Modeled Instruction

Learn

Probe students' understanding of the problem's context by asking questions similar to the following.

- Which line segments do you need to find the lengths of?
- What do you know about the angles of the corners of a rectangle?

As students read the problem again, guide them to identify the words and numbers needed to solve the problem.

- How do you find the area of a rectangle?

Reread You may wish to use this Think Aloud to demonstrate how the information given can be used to solve the problem.

This problem uses a lot of geometry. It gives a relationship between the length and the width of a rectangular courtyard and also tells the area of the courtyard. It then asks for the length of the diagonal. How can I find that length? If I have just one diagonal, then the rectangle is cut into two right triangles. Then maybe I can use the Pythagorean Theorem: $leg^2 + leg^2 = hypotenuse^2$. The sides of the rectangle are the legs of the triangle, but the problem does not tell me measures of those sides. If I knew one measure, I could find the other by using the fact that the length is $\frac{4}{3}$ the width. I might be able to use the area of the rectangle to help me. I have to start somewhere, so maybe I will make a guess for the width and see what happens.

62   Unit 3

Decide What to Do

You know the shape and area of the courtyard, how its length and width are related, and that diagonals divide it into right triangles.

Ask How can I find the length of a diagonal path between two posts?

- I can use the strategy *Guess, Check, and Revise*.
- I can guess widths and lengths until I find a rectangle with the right area. Then I can use the **Pythagorean Theorem** to find the diagonal distance.

Use Your Ideas

Step 1 Make a table to organize your guesses. Guess a width and find a corresponding length and area.

Step 2 Compare your guessed area with the given area. If it does not match, make another guess. Continue until you find the correct dimensions.

w (feet)	l (feet)	Area (square feet)	Compare Areas
300	400	300 × 400 = 120,000	120,000 > 97,200
240	320	240 × 320 = 76,800	76,800 < 97,200
270	360	270 × 360 = 97,200	97,200 = 97,200

The Pythagorean Theorem states that the sum of the squares of the lengths of the legs of a right triangle equals the square of the length of the hypotenuse. $a^2 + b^2 = c^2$

Step 3 Use the correct dimensions and the Pythagorean Theorem to find the length of a diagonal path.

The width is __270__ feet. The length is __360__ feet.

$270^2 + 360^2 = c^2$
$72,900 + 129,600 = c^2$
$202,500 = c^2$
$450 = c$

The length of a diagonal path between two posts is __450__ feet.

Review Your Work

Make sure you found the length the problem asked for.

Explain Why was the starting number for the third guess greater than 240?

Sample Answer: The previous guess of 240 was too low, so the third guess should be a greater number.

97

Try

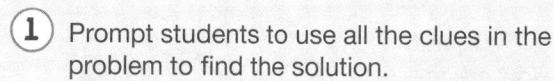

Solve the problem.

(1) A Pratt truss is a structure used in bridge building. The Pratt truss shown has eight congruent right triangles. Each triangle has a perimeter of 24 feet. The lengths of the sides are whole numbers of feet. What is the length of each side of one triangle?

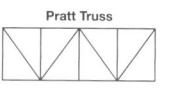

Pratt Truss

Read the Problem and Search for Information

Think about how right triangles, perimeter, and the Pythagorean Theorem are related.

Decide What to Do and Use Your Ideas

Use the strategy *Guess, Check, and Revise*. Guess lengths for each leg and the **hypotenuse** that add up to 24 feet. Check if your numbers work in the Pythagorean Theorem. If not, try again.

Ask Yourself In a right triangle, which side is always the longest?

Step 1 Organize your guesses in a table. Guess lengths for a, b, and c that add up to 24. Choose numbers so that a is the shortest side and c is the longest side.

Step 2 Substitute your guesses into the equation $a^2 + b^2 = c^2$. If they do not make the equation true, guess again.

a	b	c	$a^2 + b^2$	c^2	Does $a^2 + b^2 = c^2$?
4	8	12	80	144	80 < 144
5	9	10	106	100	106 > 100
6	8	10	100	100	100 = 100

The lengths of the sides are __6__, __8__, and __10__ feet.

Review Your Work

Check that the lengths of the sides you chose add up to 24 and that they work in the Pythagorean Theorem.

Describe Why is it helpful to use a table when making guesses?

Sample Answer: The table helps me organize my guesses so I can keep track of incorrect or repeated answers and make better guesses.

98   Unit 3   Using Algebra

Scaffolded Practice
Apply

2 Guide students to think through the steps needed to answer the question that was asked.
- Once you make a guess for a, how do you find a corresponding value for b?
- Once you know the rectangle's length and width, how can you find the length of the beam?

HOTS Conclude Students' responses should show an understanding that checking the result helps them improve their guesses.

3 Ask students about how they will choose their guesses and complete the table.
- Once you guess the length of one of the sides, how do you find the length of the third side?
- If your guess results in a sum of $a^2 + b^2$ that is too high, will you decrease the value of a or the value of c?

HOTS Appraise Students' responses should note that Guess, Check, and Revise can help them find the correct values when writing and solving equations may be difficult.

4 Help students visualize the relationship between the triangles and the rectangle.
- How could you sketch and label a diagram of the mural?
- How are the lengths of the legs of the right triangles related to the dimensions of the rectangle?

HOTS Discuss Responses should show that students know how the perimeter is related to the length and height of the mural.

5 Help students think about how to make a first guess.
- Will the lengths of sides a and b be greater than or less than the length of a rafter?
- What should be the product of a and b?

HOTS Infer Responses should indicate that students understand perimeter and the meaning of congruent figures.

Apply
Solve the problems.

2 A rectangular ceiling has an area of 168 square feet and a perimeter of 62 feet. It is strengthened by a diagonal beam. What are the lengths of the sides of the ceiling (a and b) and the beam (c)?

a	b	a × b	Does a × b = 168?
5	26	130	130 < 168; too small
11	20	220	220 > 168; too large
7	24	168	168 = 168 ✓

$a^2 + b^2 = c^2$, so __49__ + __576__ = __625__

$\sqrt{c^2} = \sqrt{625} =$ __25__

The sides are 7 feet and 24 feet, and the beam is
Answer 25 feet.

Ask Yourself: If the ceiling has a perimeter of 62 feet, what is the combined length of a and b?

Hint The product of the lengths of the sides must equal the given area of the rectangle.

Conclude Why should you check your work before making your next guess?

Sample Answer: Checking my work shows me if I should stop or how I should make my next guess better.

3 The front face of an art museum is a right triangle with a perimeter of 180 meters. The length of b is 60 meters. What are the lengths of a and c?

Art Museum Front

Hint The lengths of sides a, b, and c must satisfy $a^2 + b^2 = c^2$.

a	b	c	$a^2 + b^2$	c^2	Does $a^2 + b^2 = c^2$?
50	60	70	6,100	4,900	6,100 > 4,900; too large
40	60	80	5,200	6,400	5,200 < 6,400; too small
45	60	75	5,625	5,625	5,625 = 5,625; ✓

Answer Side a is 45 meters and side c is 75 meters.

Ask Yourself: Do my guesses for the lengths of the sides add up to 180 meters?

Appraise Why is Guess, Check, and Revise a useful strategy here?

Sample Answer: I can guess lengths that add up to 180 meters until I get lengths that make a right triangle.

Lesson 11 Strategy Focus: Guess, Check, and Revise 99

---

4 Curtis painted a mural. The mural is a rectangle made up of 8 congruent right triangles. Each triangle is painted a different color. The perimeter of the mural is 136 feet and its area is 480 square feet. The length of the mural is greater than its height. What are the dimensions of each right triangle?

Ask Yourself: If I find the length and height of the mural, how can I use that information to find the lengths of the sides of one triangle?

Hint The long side of one right triangle is $\frac{1}{2}$ the length of the entire mural.

l	h	Area	Is Area = 480?
40	28	1,120	1,120 > 480; too large
50	18	900	900 > 480; too large
60	8	480	480 = 480; ✓

The legs of each right triangle are 15 feet and 8 feet.
Answer The hypotenuse is 17 feet.

Discuss How did you choose your first guess?

Sample Answer: I chose numbers for length and height that added to 68, which is half of the perimeter, 136.

5 The gable end of a pitched roof is made from 2 congruent right triangles. The roof shown has rafters 15 feet long. The area of the gable end of the roof is 108 square feet. Side b is longer than side a. How long is the span?

Gable End of a Pitched Roof (rafter 15 ft)

Ask Yourself: What are the factors of 108?

Hint Calculate c^2. Then compare that value with your guesses of $a^2 + b^2$.

a	b	$a^2 + b^2$	c^2	Does $a^2 + b^2 = c^2$?
6	18	360	225	360 > 225; too large
9	12	225	225	225 = 225; ✓

Answer The length of the span is 24 feet.

Infer What information in the problem indicates how you might find the perimeter of the gable end of the pitched roof?

Sample Answer: The problem says that the gable end is made from 2 congruent right triangles.

100 Unit 3 Using Algebra

64 Unit 3

Practice

Solve the problems. Show your work.

6 Jordan is a construction worker. He must make a wall brace in the shape of a right triangle with sides that are whole numbers of inches long. He will use the entire length of a 60-inch board to make the brace. The short leg of the triangle will be 10 inches long. What will the lengths of the long leg and the hypotenuse be?

Some students will use the lesson strategy; however, other strategies may be used. Accept all reasonable work leading to the correct answer.

Answer The long leg will be 24 inches and the hypotenuse will be 26 inches.

Evaluate Is it reasonable to solve this problem by guessing? Why?

Sample Answer: Yes, because I know that the lengths of the two longer sides must have a sum of 50 inches and be solutions of $10^2 + b^2 = c^2$.

7 A rectangular plaza has an area of 300 square meters. Its width is $\frac{3}{4}$ its length. Both dimensions are whole numbers of meters. A diagonal path connects two corners. How long is the diagonal path?

Some students will use the lesson strategy; however, other strategies may be used. Accept all reasonable work leading to the correct answer.

Answer The diagonal path is 25 meters long.

Compare How is this problem similar to the *Learn* problem?

Sample Answer: Both problems express the lengths of the legs of the right triangle as a ratio, give the area, and ask for diagonal lengths.

Create A gardener makes a garden in the shape of a right triangle. It has a perimeter of 48 feet. Use this information to write a problem you can solve by using the strategy *Guess, Check, and Revise*.
See teacher notes.

Lesson 11 Strategy Focus: Guess, Check, and Revise   101

Create

In this lesson, students write a problem using given information. The problem should be possible to solve using the strategy *Guess, Check, and Revise*. If students are struggling, suggest they first find the side lengths of a right triangle that has a perimeter of 48 feet, then write a problem that includes a relationship between the sides of the triangle they found.

Accept student responses that include a right triangle with a perimeter of 48 feet and that provide a correct solution.

Independent Practice
Practice

Students should be encouraged to choose any strategy to solve Problems 6 and 7, though many may prefer to use *Guess, Check, and Revise*.

6 Some students may choose to write an equation to solve the problem directly.

Sample Work

a	b	c	$a^2 + b^2$	Does $a^2 + b^2 = c^2$?
10	20	30	100 + 400	500 < 900
10	22	28	100 + 484	584 < 784
10	24	26	100 + 576	676 = 676

HOTS Evaluate Students' responses should state that the strategy works because they can use the perimeter formula to help them make guesses and use the Pythagorean Theorem to check the guesses.

7 Suggest students draw a diagram to help them visualize the problem.

Sample Work

a	b	a × b
20	15	300

$a^2 + b^2 = c^2$
$20^2 + 15^2 = c^2$
$400 + 225 = c^2$
$625 = c^2$
$25 = c$

HOTS Compare Explanations should note that both problems require students to find the length of a diagonal using the lengths of the sides of a rectangle. Also, the relationship between the lengths of the sides is given, as is the area of the rectangle.

Lesson 11   65

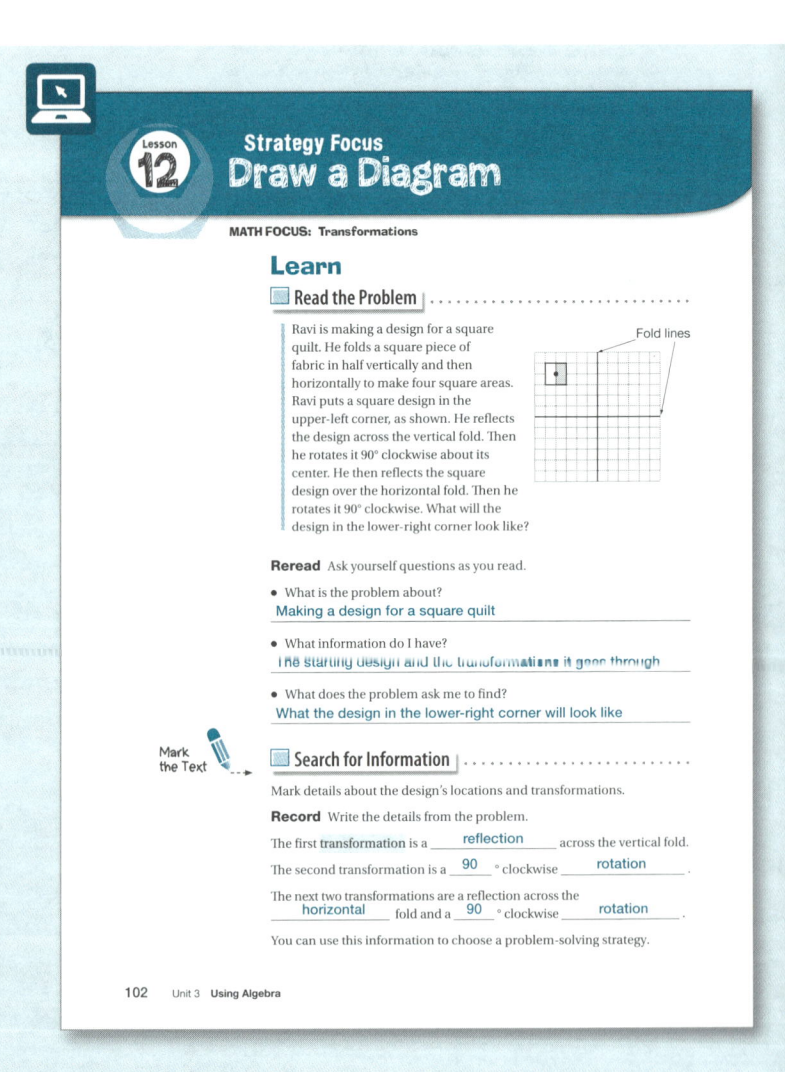

Lesson 12

Strategy Focus
Draw a Diagram

Lesson Overview

Lesson Materials: calculator

Skills to Know	Outcome	Math Vocabulary	eResources www.optionspublishing.com
• Identify and use reflections • Identify and use rotations • Identify and use translations	Students will recognize that drawing diagrams is an efficient way to solve problems involving transformations.	reflection, rotation, transformation, translation	• Interactive Whiteboard Transparency 12 • Homework, Unit 3 Lesson 12 • Coordinate Grid 2 • Problem-Solving Checklist, also available in the student worktext, page 7

Modeled Instruction

Learn

To be sure students understand the context of the problem, ask questions such as the ones below.

- How does Ravi make four square areas?
- How does Ravi start the design?

As students read the problem again, guide them to identify the words and numbers needed to solve the problem.

- What transformations does Ravi use each time he places the square design in another corner?
- Which corner of the design is the problem asking about?

Reread You may wish to use this Think Aloud to demonstrate how to choose a strategy.

This problem is about someone designing a quilt. Ravi divided the quilt into four squares and seems to be using the same design in each, except that he puts the design in a different position. The problem shows the design in the first square and tells what he does for the second and third squares. It asks what the design will look like in the third square.

I think I can solve this problem by just reading carefully. It says, He reflects the design across the vertical fold. *Now the dark portion would be on the left.* Then he rotates it 90° clockwise about its center. *Let me think about it. Does that put the dark side on the top? Then it says,* He then reflects the square design over the horizontal fold and… *I am so confused! This is way too much information to picture in my head. I think I will draw a diagram showing each step.*

66 Unit 3

Decide What to Do

You know what the design in the upper-left corner looks like. You know the transformations that will be used.

Ask How can I find out what the design in the lower-right corner will look like?

- I can use the strategy *Draw a Diagram*.
- I can use a grid to show the design. I can show the folds and the four square areas. I can draw the transformations until I reach the lower-right corner.

Use Your Ideas

Step 1 Reflect the design over the vertical fold. Perform a 90° clockwise rotation of the reflection around its center. Draw the square design in the upper-right corner.

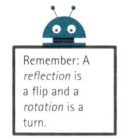

Step 2 Reflect the design over the horizontal fold. Rotate the reflection 90° clockwise around its center. Draw the square design in the lower-right corner.

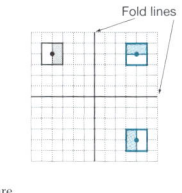

Remember: A *reflection* is a flip and a *rotation* is a turn.

So the design in the lower right corner of the quilt will look like this:

Review Your Work

Look back at the diagram. Check that you have performed each transformation correctly.

Recognize What is another way you could start with the design in the upper-right corner and end up with the design in the lower-right corner?

Sample Answer: I could slide the design in the upper-right corner over the horizontal fold. Then I could rotate it 90° counterclockwise.

103

Modeled Instruction *(continued)*

Help students make a connection between what they know and what they need to find out.

- *In what form will your answer be?*
- *How will a grid be helpful to you?*

Ask questions that guide students to consider each step in the solution process.

- *What would happen if you did not reflect the design exactly over the fold lines?*
- *How do you draw a 90° clockwise rotation?*

Emphasize to students the importance of checking that they completed each transformation in the problem correctly.

Recognize Explanations should show a transformation or a set of transformations that move the design into the lower-right corner.

Try

Solve the problem.

1. Maya designs a stencil on a coordinate plane. She draws a fish in the upper-left corner, as shown. She translates the fish 8 units down and rotates it 180° about the eye. Next, she translates that fish 6 units right. Then she translates that image 8 units up and rotates it 180° about the eye. What are the coordinates of the three points that make up the fish's mouth in the final image?

Ask Yourself: What is a translation?

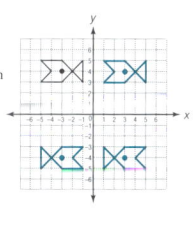

Mark the Text

Read the Problem and Search for Information

Reread the problem. Try to visualize each translation and rotation.

Decide What to Do and Use Your Ideas

You can use the strategy *Draw a Diagram*. Follow the directions to arrive at the final image of the fish. Then examine the last image.

Step 1 Translate the point that represents the eye, $(-3, 4)$, down 8 units to the point $(-3, -4)$. Draw a fish that has been rotated 180° around that point.

Step 2 Translate the eye of the second fish 6 units right to the point ___(3, -4)___. Draw the next fish around that point.

Step 3 Translate the eye of the third fish up 8 units to the point ___(3, 4)___. Draw a fish that has been rotated 180° around that point.

The coordinates of the three points that make up the fish's mouth in the final image are ___(1, 5)___, ___(2, 4)___, and ___(1, 3)___.

Review Your Work

Check that you have performed all transformations in the correct order.

Discuss Do you need the detail about the eye? Why or why not?

Sample Answer: Yes, because you can rotate a figure about any point. So you need to know about which point you should rotate the figure.

104 Unit 3 Using Algebra

Guided Practice
Try

1. Make sure students understand and follow the instructions carefully.

   Help students understand the transformations used.

   - *How many times does Maya draw the fish?*
   - *What information is given about each transformation?*

   Help students perform the transformations on a coordinate plane.

   - *How can you find the coordinates of the eye after each transformation?*
   - *Once you know the position of the eye, how do you know what direction the fish is facing?*

   Make sure that for each transformation, students have drawn the images in the correct quadrants.

   Discuss Responses should state that it is important to know the point about which the figure is rotated.

Lesson 12 67

Scaffolded Practice
Apply

2 Stress the importance of using the lines of reflection correctly.

- *When you reflect the L over the x-axis, how do you know how far down the image will be?*
- *When you reflect the L over the y-axis, how do you know how far to the left the image will be?*

HOTS Examine Responses should demonstrate an understanding that by using Al's transformations, the final L would be right side up.

3 To help students solve this problem, ask questions like these.

- *What rotation is the same as 90° counterclockwise?*
- *If you rotated the tiles before moving them, how would the result compare with your answer?*

HOTS Consider Explanations should note that the question only asks about the shaded portion of the pattern.

4 Prompt students to think about why some information is given in the problem and other information is not given.

- *Why is it not necessary for the problem to tell the direction of the 180° rotations?*

HOTS Explain Students' responses should mention that using a separate grid makes it easier to see the solution.

5 Encourage students to consider the coordinates in relation to the transformations.

- *After the translation, how do the coordinates of the vertices of the rectangles compare with those of the original rectangles?*
- *After the reflection, how do the coordinates of the vertices of the rectangles in Quadrant III compare to those in Quadrant IV?*

HOTS Analyze Explanations should show that students can identify common errors made when working with reflections.

Apply

Solve the problems.

2 Lucille takes the first letter of her name and plots it on the coordinate plane. She reflects it over the x-axis and then reflects that image over the y-axis. Draw the final image.

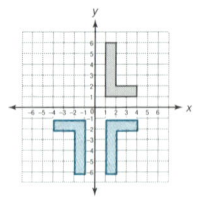

Ask Yourself Does the size of the image change with a reflection?

Hint Pick a point such as the lower-left corner of the L and plot that point first.

Answer See Quadrant III of graph.

Examine Al thinks that the same image results by translating the L down 7 units, then reflecting it over the y-axis. Is Al correct? Use the coordinates of the lower-left corner of the L to justify your answer.

Sample Answer: No, because the point (1, 1) would end up at (−1, −6). It should be at (−1, −1).

3 Tia is designing a pattern of tiles. She moves the tile shown 6 units left and rotates it 90° counterclockwise about its center. Then she moves that image 6 units up and rotates it the same way. Finally, she moves that image 6 units right and rotates it the same way one more time. What are the coordinates of the vertices of the shaded section in the last image?

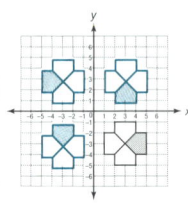

Hint Draw the outline of the figure before shading the section.

Ask Yourself Which direction is counterclockwise?

Answer The coordinates of the vertices of the shaded section in the last image are (2, 1), (2, 2), (3, 3), (4, 2), and (4, 1).

Consider Is it necessary to transform the entire figure to answer the question? Explain.

Sample Answer: No, I only need to transform the shaded part of the figure.

Lesson 12 Strategy Focus: Draw a Diagram 105

4 Kevin makes a picture of a face using three trapezoids. To make a new face, he rotates the "eyes" 180° about the points (−3, 3) and (3, 3). He rotates the "mouth" 180° about the point (0, −3). Draw the new face.

Ask Yourself Does it matter if the rotation is clockwise or counterclockwise?

Hint Draw your answer on the coordinate plane on the right.

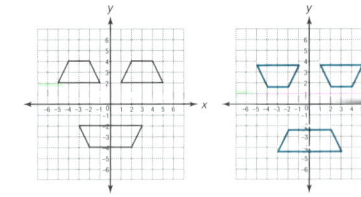

Answer See graph.

Explain Why is it helpful to use a separate grid for your answer?

Sample Answer: Putting the answer on the same grid would cause the figures to overlap.

5 Sheri is creating an asteroid field for a video game. The initial pattern is shown. She translates the entire pattern 6 units down. Then she reflects the image over the y-axis. What are the coordinates of the vertices of the largest asteroid in the final image?

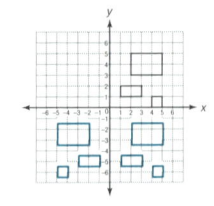

Hint Make sure you translate and reflect every rectangle the same way.

Ask Yourself Should I transform each rectangle one at a time or should I do the whole pattern all at once?

Answer The coordinates of the vertices are (−5, −1), (−2, −1), (−2, −3), and (−5, −3).

Analyze Abby says that the coordinates of the vertices are (2, 1), (2, 3), (5, 3), and (5, 1). What mistake did Abby probably make?

Sample Answer: Abby probably reflected the first image across the x-axis, not the y-axis.

106 Unit 3 Using Algebra

68 Unit 3

Practice

Solve the problems. Show your work.

6 Mike is making a sign for the 4C Construction Company. After drawing the first C, he first translates it 2 units down and 3 units right. He repeats these transformations until he has 4 Cs. What does the final sign look like?

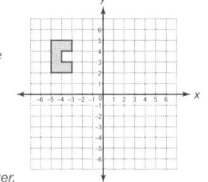

Some students will use the lesson strategy; however, other strategies may be used. Accept all reasonable work leading to the correct answer.

Answer See teacher notes.

Conclude Does it change the final sign if you do each set of transformations in the order 3 units right, then 2 units down? Explain.

Sample Answer: No, because you arrive at the same image if you go right first, then down.

7 Diana is making a design using the figure shown. For the first transformation, she reflects the figure across the y-axis and then rotates it 180° about its center. Then she reflects the image across the x-axis. Finally, for the last transformation, she reflects that image across the y-axis and then rotates it 180° about its center. Draw the final design.

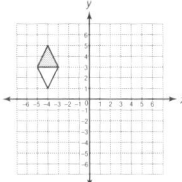

Some students will use the lesson strategy; however, other strategies may be used. Accept all reasonable work leading to the correct answer.

Answer See teacher notes.

Formulate What other question can you ask from the given information?

Sample Answer: What are the coordinates of the vertices of the final image?

Create Look back at the problems in the lesson. Choose one and change the original design and the transformations. Write a new problem that can be solved using the strategy *Draw a Diagram*. Solve your problem. See teacher notes.

Lesson 12 Strategy Focus: Draw a Diagram 107

Create In this lesson, students modify a problem from the lesson by providing their own design and describing new transformations. If students are struggling, suggest they use a simple design, such as a square with one half of it shaded.

Accept student responses that require the application of two or more transformations for the solution and that give a correct final diagram.

Independent Practice
Practice

Students should be encouraged to choose any strategy to solve Problems 6 and 7, though many may prefer to use *Draw a Diagram*.

6 Some students may choose to describe the locations of the Cs in the final sign instead of drawing a diagram.

Sample Work

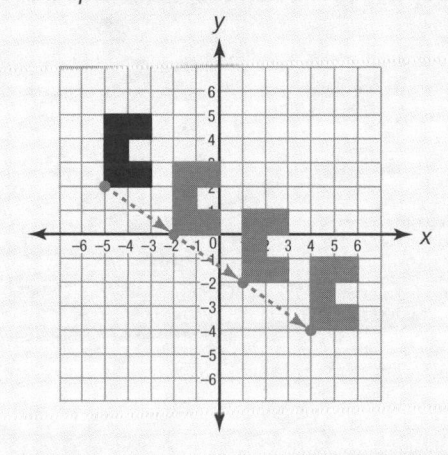

Conclude Responses should note that in this problem, the order of the translations does not affect the final image.

7 Emphasize to students the importance of following the steps in the problem precisely. Have them note which axis to use for each line of reflection.

Sample Work

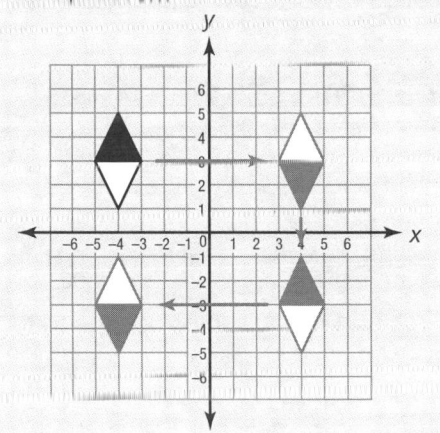

Formulate Responses should show that students understand what can be determined from the given information and resulting images.

Lesson 12 69

UNIT 3 Review

UNIT 3 Review

In this unit, you worked with four problem-solving strategies. You can often use more than one strategy to solve a problem. So if a strategy does not seem to be working, try a different one.

Problem-Solving Strategies
- Make a Graph
- Make a Table
- Guess, Check, and Revise
- Draw a Diagram

Check students' work throughout. Students' choices of strategies may vary.

Solve each problem. Show your work. Record the strategy you use.

1. A city has a rule that there must be no fewer than 1 mailbox for every 3 downtown blocks. What is the minimum number of mailboxes needed for 10 downtown blocks?

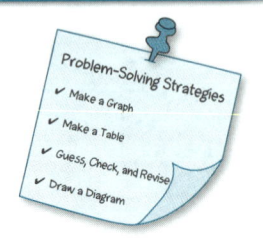

Answer: _The minimum number of mailboxes is 4._

Strategy: _Possible Strategy: Make a Graph_

2. A rectangular mural has an area of 240 square feet. Its dimensions are whole numbers of feet. Its width is $\frac{5}{12}$ its length. How long are the mural's diagonals?

Answer: _The diagonals are 26 feet long._

Strategy: _Possible Strategy: Guess, Check, and Revise_

3. A truck driver makes deliveries at the rate shown on the graph. How long will it take the driver to make 16 deliveries?

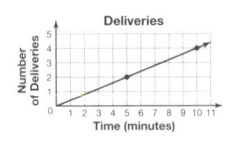

Answer: _It will take the driver 40 minutes to make 16 deliveries._

Strategy: _Possible Strategy: Make a Table_

4. Jane is making a design. She starts with the figure at the upper left. She reflects it over the *x*-axis, reflects the new image over the *y*-axis, and finally reflects that image over the *x*-axis. What does the final image look like?

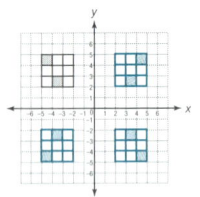

Answer: _See Quadrant I of graph._

Strategy: _Possible Strategy: Draw a Diagram_

5. A cab driver is 7 miles from the main garage. He drives toward the garage at the rate shown on the graph. How long will it take the cab to reach the garage?

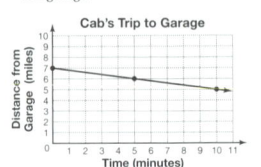

Answer: _It will take the cab 35 minutes to reach the garage._

Strategy: _Possible Strategy: Make a Table_

Explain how you found your answer.

Sample Answer: I found the slope of the graph, which gave me the rate of $\frac{1}{5}$ mile per minute. I used the rate to make a table in order to extend the pattern.

108 Unit 3 Using Proportional Reasoning

109

Support for Assessment

The problems on pages 108–111 reflect strategies and mathematics students used in the unit.

Although this unit focuses on four problem-solving strategies, students may use more than one strategy to solve the problems or use strategies different from the focus strategies. Provide additional support for those students who need it.

Make a Graph For Problems 1 and 9, remind students to find at least 2 data points from the information given. If the answer students get is not a whole number, have students reread the problem to check how they should round.

Make a Table For Problems 3, 5, and 6, elicit that a table can be helpful to students in extending the patterns shown in the graphs. Help students understand the contexts of the problems. For Problem 5, ask, *Why does the distance decrease as the time increases?* For Problem 6, ask, *Why is the first point on the graph (0, 4)?*

Guess, Check, and Revise For Problems 2, 8, and 10, have students describe how they would make a first guess. Once they find the number(s) that work, have them read the question again to make sure they solved the problem completely.

Draw a Diagram For Problems 4 and 7, suggest that students draw one step of the problem and check to see that they followed the instructions correctly before they draw the next step.

You may wish to use the *Review* to assess student progress or as a comprehensive review of the unit.

Promoting 21st Century Skills

Write About It
Communication

When students are asked to find an alternate process, they must be flexible in their thinking as they attempt to look at the problem in a different way. Students' responses should list the steps of the new transformations clearly and completely.

Team Project: Plan a Garden
Collaboration: 3–4 students

Review the expectation that in a group project, everyone's ideas should be considered and everyone must help with the work. Students should read the plan carefully and brainstorm ideas. They should then have time to decide on their own preferences before they try coming to a consensus about the number and sizes of the garden beds and the types and numbers of plants they want to include.

Ask questions that help students summarize their thinking. *How did you decide which plants fit in which garden bed? How did you keep track of your decisions?*

Extend the Learning
Media Literacy

If you have Internet access, navigate to sites where students can find growing requirements for the vegetables they would like to grow.

Search: vegetable space requirements

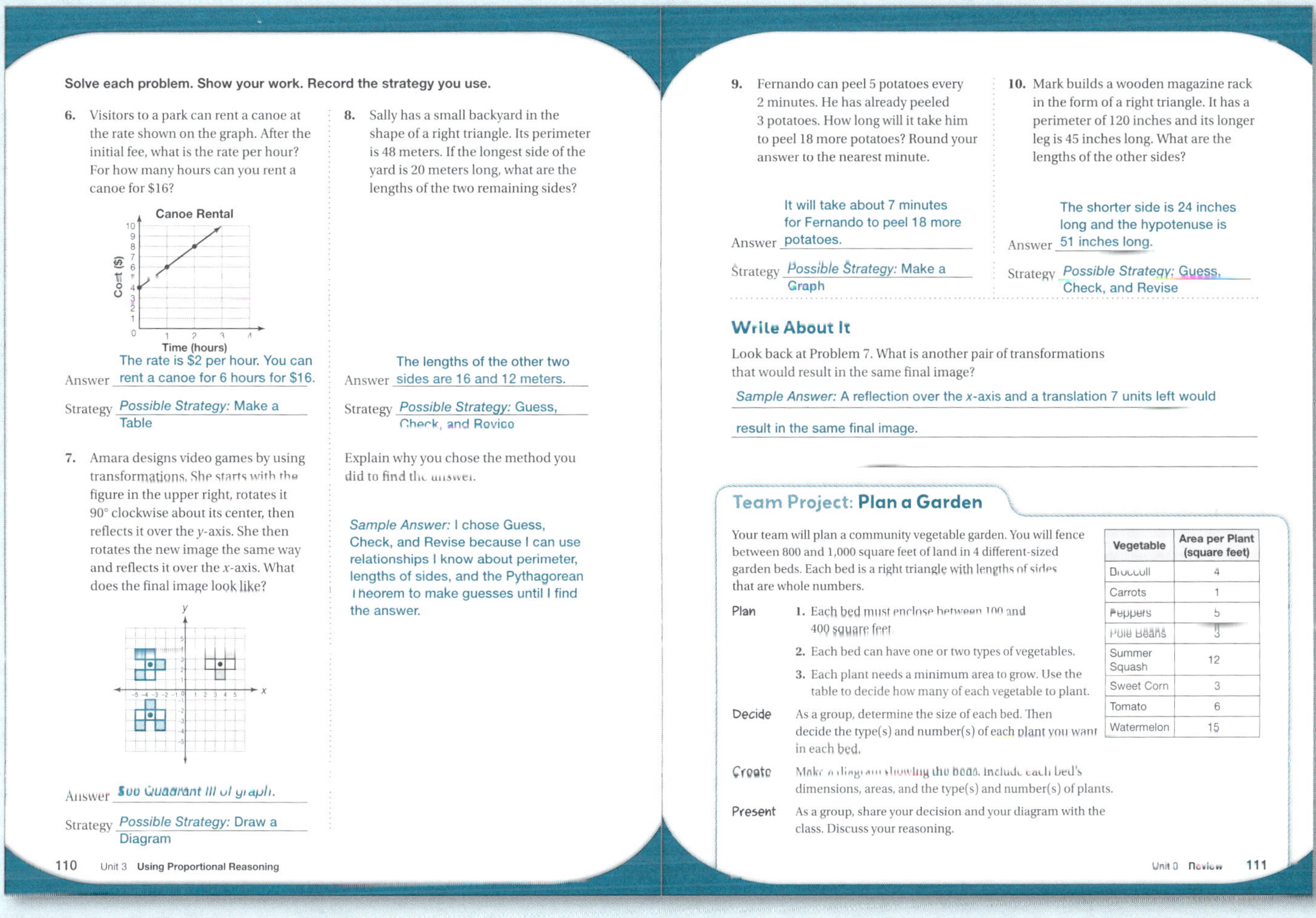

Unit 3 71

UNIT 4 Problem Solving Using Geometry

CCSS 8.G Geometry

Unit Overview

Lesson	Problem-Solving Strategy	Math Focus
13	Use Logical Reasoning	Congruent Figures
14	Look for a Pattern	Similarity and Patterns
15	Draw a Diagram	Dilations
16	Look for a Pattern	Symmetry and Tessellations

Promoting Critical Thinking

Higher order thinking questions occur throughout the unit and are identified by this icon: HOTS. These questions progress through the cognitive processes of remembering, understanding, applying, analyzing, evaluating, and creating to engage students at all levels of critical thinking.

UNIT 4 Problem Solving Using Geometry

Unit Theme: Creative Activities

Math plays an important role in creative activities. These activities can range from designing artistic tile mosaics, to planning marching band formations, to making patchwork quilts. In this unit, you will see how math and creativity often go hand in hand.

Math to Know
In this unit, you will apply these math skills:
- Use attributes of plane figures
- Use similarity in plane figures
- Use symmetry and tessellations

Problem-Solving Strategies
- Use Logical Reasoning
- Look for a Pattern
- Draw a Diagram

Link to the Theme

Write another paragraph about the artwork. What details could Sophie include in her description? Use words that describe shapes and sizes.

Sophie's class is on a field trip to the art museum. She has to write about one piece of art that she likes. She chooses this one.

Students' paragraphs will vary, but should include some words that describe shapes and sizes.

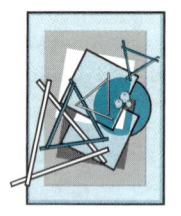

Use Math Language

Review Vocabulary

The list below shows vocabulary terms in this unit. Knowing the meaning of these terms will help you understand the problems.

dilation	equilateral	scale	symmetry
enlarged	quadrilateral	scale factor	tessellation

Vocabulary Activity   Word Parts

Parts of a word such as a prefix or a suffix can give you a clue to a word's meaning. Use words from the list above to complete the following sentences.

1. The prefix *equ-* means same or equal. An ___equilateral___ triangle has three equal sides.
2. The prefix *quad-* is related to the number four. A ___quadrilateral___ is a polygon that has four sides.
3. The suffix *-tion* means the result of. A ___tessellation___ is the result of shapes being fitted together in a pattern that contains no gaps or overlaps.
4. The result of resizing a geometric figure is called a ___dilation___.

Graphic Organizer   Word Map

Complete the graphic organizer.
- Write a definition of *symmetry*.
- Use the term in a sentence.
- Show an example of the term.
- Show a non-example of the term.

Sample Answers:

Definition	Sentence
a property of a figure that allows it to look the same after a rotation or reflection	I know a rectangle has line symmetry because if I draw a line down the middle, the two halves match.
Example	Non-Example

symmetry

Link to the Theme   Creative Activities

Ask students to read the direction line and story starter. If students are having trouble getting started, ask questions such as, *What might Sophie like about this piece of art? What geometric figures do you see in it?*

Unit 4 Differentiated Instruction

Extra Support

Some students may need reinforcement of geometry concepts.

Triangles and Quadrilaterals Provide students with examples of different types of triangles and quadrilaterals. Ask pairs of students to describe the figures to each other. Guide students to name each figure as specifically as possible.

Ordered Pairs Demonstrate how to use ordered pairs to name the locations of points in each quadrant. Give coordinate grids with 6 to 8 points plotted to pairs of students. Have students take turns naming the ordered pair for each point.

Challenge Early Finishers

Early finishers may enjoy the challenge of solving open problems with multiple solutions. You can make many problems in this program or your basal program open by removing at least one piece of information that would restrict the answer.

Closed Problem: One Solution
Allison wants to enlarge a design that is in the shape of a rectangle. The coordinates of the vertices of the design are $A(0, 0)$, $B(0, 4)$, $C(6, 4)$, and $D(6, 0)$. She plots points for the enlarged design at $A'(0, 0)$, $B'(0, 8)$, and $C'(12, 8)$. Where should she plot point D' so the enlarged design is not distorted?

Open Problem: Multiple Solutions
Allison wants to enlarge a design that is in the shape of a rectangle. The coordinates of the vertices of the design are $A(0, 0)$, $B(0, 4)$, $C(6, 4)$, and $D(6, 0)$. She plots a point for the enlarged design at $A'(0, 0)$. Where can she plot the other points so the enlarged design is not distorted?

By removing the points B' and C', students are given the opportunity to demonstrate their understanding of dilations and enlargements with their own examples. Open problems provide opportunities for students to hone their higher-level thinking skills.

English Language Learners

Build Background

Scale Models Several problems in this unit refer to scale models. Explain that a scale model is a smaller or larger version of an object that has the same proportions, or relative measurements, as the actual object. Builders often make a smaller scale model before building the larger actual object. Find pictures on the Internet of scale models of buildings or other objects and show them to students. Ask students why scale models are helpful.

Writing

Adjective to Adverb Write the word *correctly* on the board and circle *-ly*. Explain that *-ly* is a suffix, a word part added to the end of a word that changes the meaning of the word. The suffix *-ly* can change adjectives into adverbs. For example, the adjective *correct* can be used to describe a noun, while the adverb *correctly* can be used to describe a verb. Write the following frames on the board, one beneath the other:
Clara used the _____ measurements.
Clara used the measurements _____.
Guide students to copy the frames and fill in the blanks with *correct* and *correctly*. Discuss with students that *correct* describes the noun *measurements*, and *correctly* describes the verb *used*. Have students write their own sentences using other *-ly* adverbs, such as *differently*, *accurately*, and *carefully*.

Listening and Speaking

Changes in Size This unit has many problems involving changes to the size of objects. To be sure students understand the terms used to describe these changes, write *reduce* and *enlarge* on the board. Explain that *reduce* means to make smaller, and *enlarge* means to make larger. Draw a square on the board and tell students to suppose it is a photo. Ask, *What happens when I reduce this photo?* Then ask, *What happens when I enlarge this photo?*

Lesson 13
Strategy Focus
Use Logical Reasoning

Lesson Overview

Lesson Materials: ruler, calculator

Skills to Know	Outcome	Math Vocabulary	eResources www.optionspublishing.com
• Identify and use attributes of polygons	Students will recognize that using logical reasoning is an efficient way to solve some problems that involve using the attributes of polygons.	equilateral, quadrilateral	• Interactive Whiteboard Transparency 13 • Homework, Unit 4 Lesson 13 • Know-Find Table • Problem-Solving Checklist, also available in the student worktext, page 7

Modeled Instruction
Learn

Ask questions to confirm students' comprehension of the problem's context.

- *What does it mean for triangles to be congruent?*
- *What do you think it means when Ariel's mother tells her that it is not possible to make the decoration as planned?*

As students read the problem again, pose questions to help them recognize important facts.

- *What kind of information does the plan give you?*
- *How do you know what the measures of ∠FEA and ∠BCA are?*

Use a Graphic Organizer You may wish to use this Think Aloud to demonstrate how to identify information needed to solve the problem and how that information can be used.

I know the problem is about a decoration with congruent triangles. The information about the plan Ariel drew for this decoration is given in the drawing. I am going to study the drawing, filling in the Know-Find Table to help me organize the details and see how I might use them to solve the problem. In the Know column, I can put the measures of the angles. I will write: m∠AFE = 50°, m∠ABC = 60°, m∠FEA = 90°, and m∠BCA = 90°. In the Find column, I can write the question I need to answer, Is it possible to make the decoration exactly as planned?

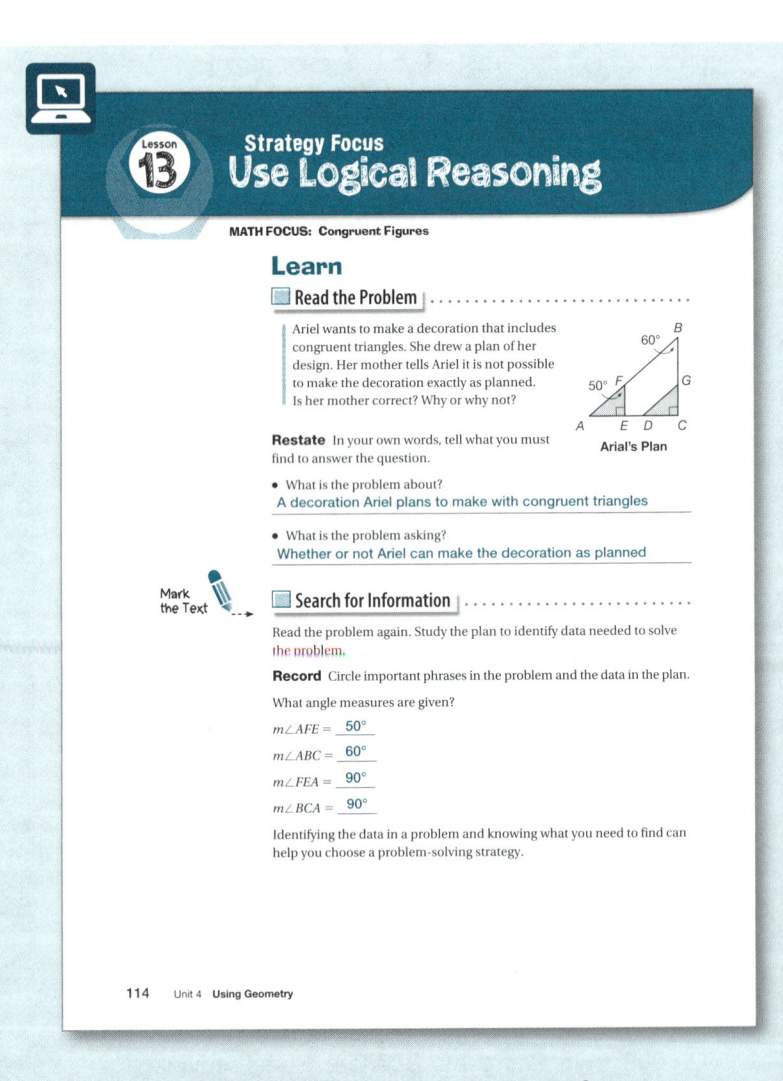

Lesson 13
Strategy Focus
Use Logical Reasoning

MATH FOCUS: Congruent Figures

Learn

Read the Problem

Ariel wants to make a decoration that includes congruent triangles. She drew a plan of her design. Her mother tells Ariel it is not possible to make the decoration exactly as planned. Is her mother correct? Why or why not?

Arial's Plan

Restate In your own words, tell what you must find to answer the question.

- What is the problem about?
  A decoration Ariel plans to make with congruent triangles
- What is the problem asking?
  Whether or not Ariel can make the decoration as planned

Mark the Text

Search for Information

Read the problem again. Study the plan to identify data needed to solve the problem.

Record Circle important phrases in the problem and the data in the plan.

What angle measures are given?

m∠AFE = **50°**

m∠ABC = **60°**

m∠FEA = **90°**

m∠BCA = **90°**

Identifying the data in a problem and knowing what you need to find can help you choose a problem-solving strategy.

114 Unit 4 Using Geometry

I know that the sum of the angle measures of a triangle is always 180°. I wonder if I can use this fact and the angle measures I know to find the measures of ∠EAF and ∠CAB. They are the same. So, if the measures are the same, the drawing makes sense.

74   Unit 4

Decide What to Do

You know the shapes in the decoration and the measures of some angles. You also know that the plan may not be drawn correctly.

Ask How can I find out if something in the plan is incorrect?

- I can use geometric relationships to find measurements that are not given in the plan.
- I can *Use Logical Reasoning* to see if anything in the drawing does not make sense.

Use Your Ideas

Step 1 Use geometric relationships to find the measure of ∠CAB.

$m\angle ABC + m\angle BCA + m\angle CAB = \underline{180°}$

$\underline{60°} + \underline{90°} + m\angle CAB = \underline{180°}$

$m\angle CAB = \underline{30°}$

The sum of the measures of the angles of a triangle is always 180°.

Step 2 Use geometric relationships to find the measure of ∠EAF.

$m\angle AFE + m\angle FEA + m\angle EAF = \underline{180°}$

$\underline{50°} + \underline{90°} + m\angle EAF = \underline{180°}$

$m\angle EAF = \underline{40°}$

Step 3 Compare the results of Step 1 and Step 2 and use logical reasoning to make a conclusion.

$m\angle CAB \neq m\angle EAF$

But ∠CAB and ∠EAF name the same angle.

Ariel's mother is correct. It is not possible that the same angle could measure __30__° and also measure __40__°.

Review Your Work

Check your reasoning and calculations. Did you apply the geometric relationships correctly?

Explain How did you use logical reasoning to solve this problem?

Sample Answer: I knew that the sum of the measures of the angles in a triangle is always 180°. That isn't true in Ariel's plan.

115

Modeled Instruction (continued)

Help students make a connection between what they know and what strategy they can use to solve the problem.

- *How does the information about the triangles suggest an attribute that you can use?*
- *How can logical reasoning help you answer the question the problem asks?*

Pose questions that give meaning to each step in the solution process.

- *Why is it important to have found out that $m\angle CAB$ and $m\angle EAF$ are not equal?*

Emphasize the importance of checking that calculations made while applying geometric relationships are correct.

HOTS Explain Students' explanations should demonstrate an understanding of properties of triangles and the reasoning they used to solve the problem.

Try

Solve the problem.

① Marcos is planning a tile mosaic. He wants to use some square tiles. He checks the labels on two packs of tiles. Which pack should he buy to be sure to get some squares? How do you know?

Puzzle Pack
All the tiles are four-sided and equilateral. Some of the tiles have four right angles.

Mystery Pack
All the tiles are quadrilaterals. All squares have four sides.

Mark the text

Read the Problem and Search for Information

Circle the data in the problem that will help you solve it.

Decide What to Do and Use Your Ideas

Use Logical Reasoning to reach a conclusion.

Step 1 Understand the vocabulary.

A **quadrilateral** is a polygon with __4__ sides.

Equilateral means having sides of __equal__ length.

Ask Yourself: Do all the tiles in the pack have to be squares?

Step 2 Use logic to evaluate the labels.

Puzzle Pack: All tiles have __4__ sides of equal __length__.

Some of those tiles have __4 right__ angles.

So some of those tiles must be __squares__.

Mystery Pack: All the tiles are __quadrilaterals__.

None of the tiles have to be __squares__.

Marcos should buy the __Puzzle Pack__ because it must have some squares.

Review Your Work

To be sure of your answer, try to find a way for each pack to contain no squares.

Identify How can you change one label so that both packs must have some squares?

Sample Answer: Change the second line on the Mystery Pack label to: "Some of the tiles have 4 congruent sides and 4 right angles."

116 Unit 4 Using Geometry

Guided Practice

Try

① Encourage students to apply what they know about quadrilaterals and squares as they use logical reasoning to solve this problem.

Discuss the context of the problem as well as the information needed to solve it.

- *Which sentences on the labels give information about the shapes of the tiles in the packs?*

Ask students how they know what is in each pack.

- *What kinds of quadrilaterals are equilateral?*
- *What kinds of quadrilaterals have four right angles?*

Have students draw or name a polygon that is four-sided and equilateral but is not a square.

- *Can a four-sided equilateral tile have four right angles and not be a square?*

HOTS Identify Responses should indicate changing the label for *Mystery Pack* to say it contains some squares.

Lesson 13   75

Scaffolded Practice
Apply

2 Make sure students recall that the sum of the lengths of any two sides of a triangle must be greater than the length of the third side.

- How does drawing $\overline{EB}$ help to solve this problem?
- What does the diagram tell you about the length of $\overline{EB}$?

HOTS Determine Students' explanations should mention the importance of checking a diagram that looks reasonable to see if indeed it is.

3 Prompt students to think about all the properties of equilateral triangles.

- How does knowing the angle measures of some of the tiles in the *Mosaic Mix* help you know whether some of the tiles are equilateral triangles?

HOTS Differentiate Responses should demonstrate that students understand that shapes other than triangles can be equilateral as long as their sides are all the same length.

4 Make sure students recall the attributes of a parallelogram.

- How are the angles of a parallelogram related?

HOTS Analyze Responses should show that students understand that the sum of the measures of the angles of any quadrilateral is 360°.

5 Ask students to identify which sentences on the labels refer to what is inside the packs and which are general facts.

- How did you decide what shapes could be in the *Polygon Assortment*? In the *Quad Pack*?

HOTS Assess Responses should show that students know that all squares are rectangles.

Apply

Solve the problems.

2 Whitney draws a plan for a plaque. Don says the dimensions she is using are not possible. Whitney does not believe him. What facts can Don use to show he is correct?

Whitney's Plan

Quadrilateral BCDE is a ___rectangle___. So EB = ___25 inches___.

In △ABE, the measure of $\overline{EB}$ must be greater than ___9 inches___ and less than ___23 inches___.

The facts Don can use are that EB must be 25 inches long to make the rectangle, but that same segment must be less than 23 inches long to make the triangle.

Answer Both facts cannot be true.

Hint: Draw a line segment from E to B.

Ask Yourself: Can both of these facts be true?

Determine Why is it helpful to use logical reasoning for this problem?

Sample Answer: The plan looks convincing, but using logical reasoning proves that a plaque with these dimensions and this shape is impossible.

3 Richard is creating a mosaic out of tiles. He has a choice of two packs. Which of the packs should he buy so he is sure to get some equilateral triangle tiles? Explain how you know.

Equi-Tiles — All the tiles are equilateral. Some triangles are equilateral.
Mosaic Mix — All the tiles are triangles. Some of the tiles have angles that are all equal to 60°.

The angles in an equilateral triangle measure ___60°___, ___60°___, and ___60°___.

___Equilateral___ means having sides that are all the same length.

Answer Richard should buy Mosaic Mix, because triangles with three 60° angles must be equilateral.

Hint: Use both facts in each pack to solve the problem.

Ask Yourself: Is a triangle the only shape that can be equilateral?

Differentiate Which tile shapes are definitely in the Equi-Tiles pack?

Sample Answer: The pack could have an equilateral triangle, a square, or a rhombus. No one shape is definitely in the pack.

Lesson 13 Strategy Focus: Use Logical Reasoning 117

---

4 Joy is making parallelogram-shaped tiles in ceramics class. She plans to make the acute angles 50° and the obtuse angles 150°. Will she be able to make her tiles exactly as planned? Explain.

Hint: Draw and label a diagram.

Ask Yourself: What must be the measures of the other 2 angles in Joy's plan?

The sum of the measures of the angles in the tiles that Joy plans to make would be ___400°___.

The sum of the measures of the angles in a quadrilateral must be ___360°___.

Answer No, because the angle measures of the tiles in Joy's plan add up to 400°. They have to add up to 360°.

Analyze How could Joy change the measures of the angles in her plan so she could make the tiles?

Sample Answer: Joy could make the angles 50° and 130° or 30° and 150°.

5 Tom is making a collage with paper pieces. He has plenty of square pieces, but he would like to buy a pack of shapes that contains no squares. Which pack should he buy? Explain.

Polygon Assortment — No piece is equilateral. All squares are equilateral.
Quad Pack — All the pieces are quadrilaterals. Some of the pieces are rectangles.

Hint: Think about the meaning of each statement on the labels.

Ask Yourself: If the pack had a square in it, could the statements on the label be true?

What figures could be in the Polygon Assortment?

Sample Answer: rectangles, trapezoids, right triangles

What figures could be in the Quad Pack?

Sample Answer: rectangles, trapezoids, squares, rhombuses

Answer Tom should buy the Polygon Assortment. Because it has no equilateral pieces, it cannot include squares.

Assess What information that was not given in the problem did you use to solve this? Give an example.

Sample Answer: I used what I knew about geometric shapes. For example, all squares are both quadrilaterals and rectangles.

Practice

Solve the problems. Show your work.

6 In wood shop, Diane wants to make a tray in the shape of a right triangle. She plans for one of the angles to be 100°. Will she able to make the tray exactly as planned? Explain.

Some students will use the lesson strategy; however, other strategies may be used. Accept all reasonable work leading to the correct answer.

Answer No, if a triangle had a 100° angle and a right angle, the total of the angle measures would be over 180°.

Compare How is Problem 4 similar to this problem?

Sample Answer: Both have impossible designs described in words. The sum of the measures of the angles is incorrect for both designs.

7 Owen is filling a jar with colorful tiles. For variety, he wants to include tiles shaped like triangles and quadrilaterals. Which pack of tiles should he buy? Explain how you know.

Brilliant Tiles
All tiles are equilateral. Some triangles are equilateral.

Pretty Polygon Tiles
Some tiles are rectangles. The measures of the angles of some of the tiles add up to 180°.

Some students will use the lesson strategy; however, other strategies may be used. Accept all reasonable work leading to the correct answer.

Answer Owen should buy Pretty Polygon Tiles, because rectangles are quadrilaterals and triangles are the only polygons with angle measures that add up to 180°.

Conclude Which pack may have only triangles? Explain.

Sample Answer: Brilliant Tiles, because all the pieces could be equilateral triangles. It can't be Pretty Polygon Tiles because the label says some tiles are rectangles.

Create Write a problem about a tile maker who is making a triangular tile. Include a plan for a triangle that cannot be made exactly as planned. Write the problem so it can be solved using logical reasoning. Solve your problem.
See teacher notes.

Lesson 13 Use Logical Reasoning 119

Create

In this lesson, students solved problems using the attributes of polygons to reason logically. Now they must write their own problem about an impossible triangle. If students are struggling, suggest they start by drawing a possible design with angle measures whose sum is 180°, then change one of the numbers.

Accept student responses that include a plan that is impossible because of its angle or side measurements. Solutions should include the reasoning used to answer the question the problem asks.

Independent Practice
Practice

Students should be encouraged to choose any strategy to solve Problems 6 and 7, though many may prefer to use *Use Logical Reasoning*.

6 Some students may draw a diagram to help answer the question.

Sample Work

The sum of the measures of the angles of a triangle is 180°.

A right angle measures 90°.

So the sum of the measures of two of the angles of Diane's planned triangle is 90° + 100°, or 190°.

190° > 180°

Compare Responses should indicate that the contexts of the problems are similar—both reflect plans that are physically impossible to execute because they do not conform to known attributes of polygons.

7 Encourage students to read the labels carefully and think about the attributes of triangles and quadrilaterals.

Sample Work

Brilliant Tiles: Equilateral triangles and some equilateral quadrilaterals (squares and rhombuses) could be in this pack, but might not be.

Pretty Polygons: Rectangles are quadrilaterals, so quadrilaterals are in this pack. Polygons with angles whose sum is 180° are triangles, so triangles are also in this pack.

Conclude Students' explanations should demonstrate the ability to reason logically to interpret the clues to show that *Brilliant Tiles* may have only triangles.

Lesson 13 77

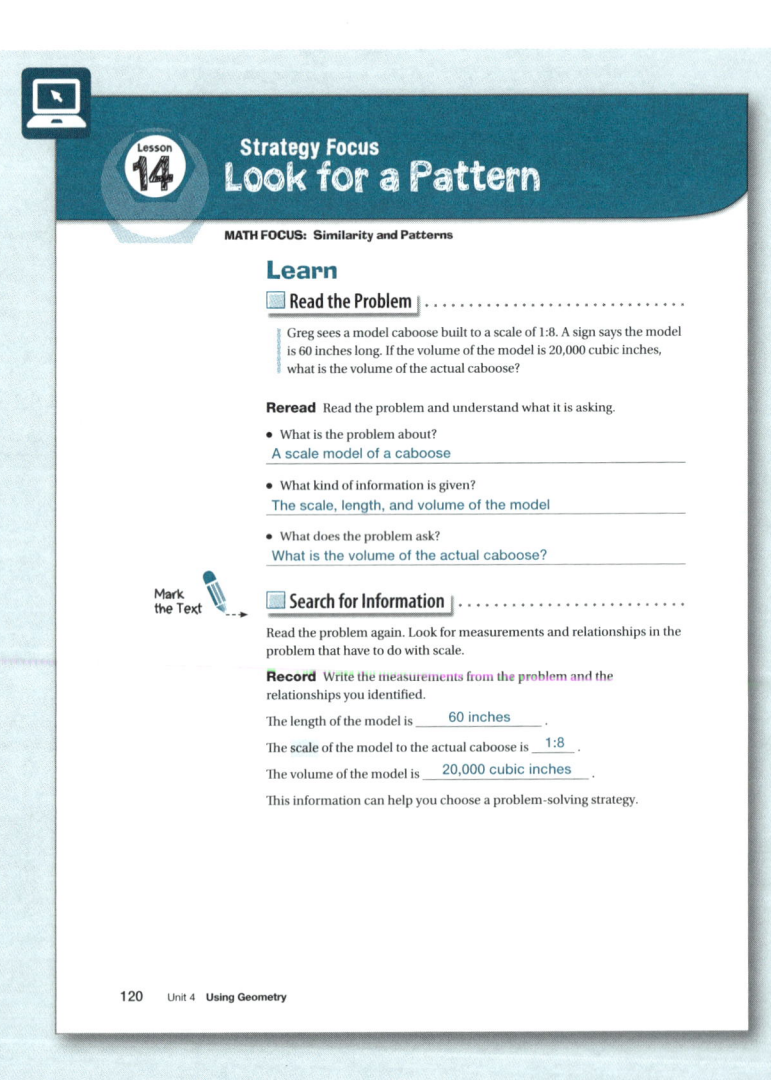

Lesson 14
Strategy Focus
Look for a Pattern

Lesson Overview

Lesson Materials: calculator

Skills to Know	Outcome	Math Vocabulary	eResources www.optionspublishing.com
• Identify and continue numerical patterns	Students will recognize that looking for a pattern is an efficient way to solve problems that involve scale models and geometric patterns.	scale	• Interactive Whiteboard Transparency 14 • Homework, Unit 4 Lesson 14 • Problem-Solving Checklist, also available in the student worktext, page 7

Modeled Instruction
Learn

To be sure students understand the context of the problem, ask questions like the ones below.

- *Is the model caboose larger than or smaller than the actual caboose? How do you know?*
- *What is volume?*

As students read the problem again, guide them to identify the words and numbers needed to solve the problem.

- *What does a scale of 1:8 mean?*
- *Do you need to know the length of the model to answer the question? Why or why not?*

Reread You may wish to use this Think Aloud to demonstrate how to read a problem to decide upon a problem-solving strategy.

I know the problem is about a scale model of a caboose and the actual caboose. The problem tells the volume of the model and asks me to find the volume of the actual caboose. The problem does not give me enough dimensions to find the volume directly. I know that the scale is 1:8, though. So I can multiply the volume of the model by 8 to find the volume of the actual caboose.

Wait, that is not right. The scale means that 1 unit of length in the model represents 8 units of length in the actual caboose. But units of volume are different than units of length, so I need to multiply by something else. Maybe I can look for a pattern in the way volume changes when I multiply dimensions by different numbers to find out how to solve this problem.

78 Unit 4

Decide What to Do

You know the length and volume of the model caboose and its scale. You also know that the measurements are proportional.

Ask How can I find the volume of the actual caboose?

- I can use the strategy *Look for a Pattern* to see how a change in volume is related to a change in dimensions.
- I can use simple examples to help me find a pattern that I can apply to solve the problem.

You need to find how the volume of the actual caboose relates to the volume of the scale model.

Use Your Ideas

Step 1 Explore what happens to the volume of a cube when you change its dimensions. If the length is multiplied by 2, the volume is multiplied by 2^3. Complete the table to find a pattern.

Length of an Edge (in.)	1	2	3	4
Volume (in.³)	1 × 1 × 1	2 × 2 × 2	3 × 3 × 3	4 × 4 × 4

The table shows that when you multiply the dimensions of a cube by a factor, its volume is multiplied by that same factor to the third power. This applies to any 3-dimensional figure.

Step 2 Use the pattern. The scale of the model to the actual caboose is 1:8. If the length is multiplied by 8, the volume is multiplied by 8^3.

Volume of actual caboose = volume of model × 8^3
= 20,000 cubic inches × 512
= 10,240,000 cubic inches

So the volume of the actual caboose is 10,240,000 cubic inches.

Review Your Work

Check that your calculations are correct.

Recognize How can you solve this problem knowing only the scale of the model to the actual caboose?

Sample Answer: I can solve this problem by finding a pattern in the relationship between the length of an edge and the volume of a cube.

Modeled Instruction (continued)

Help students make a connection between what they know and what they need to find out.

- *Can you use the scale and the length of the model to find the volume directly? Explain.*
- *Why will looking at simple examples help you solve the problem?*

Pose questions that help students focus on the steps used to solve the problem.

- *How do you find the volume of a cube?*
- *Why does this pattern allow you to find the volume of the actual caboose even though the caboose is not a cube?*

Emphasize the importance of checking that calculations are correct when using large numbers.

- *How can you use estimation to check your answer?*

Recognize Explanations should indicate that if each dimension of a solid figure is multiplied by the same factor, the volume is multiplied by that factor 3 times.

Try

Solve the problem.

1 Kay is building a pyramid with cube blocks. The top three layers are shown. How many blocks will she need for the bottom layer if she makes the pyramid 8 layers tall?

Read the Problem and Search for Information

Identify what the problem asks you to find. Study the diagram, and then circle important data in the problem.

Decide What to Do and Use Your Ideas

You can use the strategy *Look for a Pattern* to predict how many blocks the 8th layer will contain.

Step 1 Identify the pattern. Complete the table.

Layer	1	2	3	4	5	6	7	8
Number of Blocks	1 × 1	3 × 3	5 × 5	7 × 7	9 × 9	11 × 11	13 × 13	15 × 15

Ask Yourself How do the length and width of each layer change from one layer to the next?

Step 2 Apply the pattern.
15 × 15 = 225

Kay will need 225 blocks for the bottom layer.

Review Your Work

Check that there is a pattern in the number of blocks for each layer.

Identify What pattern did you find to help you solve the problem?

Sample Answer: The number of blocks in each layer is a square number. The first layer was 1 by 1, the next layer 3 by 3, then 5 by 5. The dimensions were consecutive odd numbers.

Guided Practice

Try

1 Prompt students to look for a pattern.

Discuss the context of the problem as well as the information needed to solve it.

- *How many blocks are in the bottom layer of a 2-layer pyramid? A 3-layer pyramid?*

Have students explain how they completed the table.

- *How do you know how many blocks will be on a side of the bottom of a 4-layer pyramid?*

Have students check that their pattern works for every number in the table.

Identify Responses should show that students understand both aspects of the pattern they found.

Lesson 14 79

Scaffolded Practice
Apply

② Ask students how this problem is like the *Learn* problem and how they will use a similar method to solve it.

- How will you use the scale to solve the problem?

HOTS Determine Students' explanations should show an understanding of the effect on the volume of a solid figure when all of the dimensions are multiplied by the same factor.

③ Prompt students to state the pattern to ensure they can complete the table correctly.

- How is each frame different from the one before it?

HOTS Interpret Responses should show that students understand that the table does not give the final answer. They must use the pattern they found to answer the question.

④ Encourage students to compare this problem to Problem 2 and predict the pattern they will use to find the answer.

- How is this problem like Problem 2? How is it different?

HOTS Apply Responses should show that students understand the relationship between the change in length and the change in area.

⑤ Make sure students understand what the problem asks them to find.

- What do you need to do after you complete the table to find the answer?

HOTS Sequence Responses should show that students can recognize other types of patterns that can be used to solve problems efficiently.

80   Unit 4

Apply

Solve the problems.

② April has a model of a cedar chest. Its volume is 4,096 cubic inches. April wants to make a full-size cedar chest with dimensions four times as great. What will be the volume of the full-size cedar chest?

Ask Yourself: What scale can I use to show that April's cedar chest will have dimensions 4 times as great as the model?

Length of Edge of Cube (in.)	1	2	3	4
Volume of Cube (in.³)	1 × 1 × 1	2 × 2 × 2	3 × 3 × 3	4 × 4 × 4

If the length is multiplied by 2, the volume is multiplied by 2^3.
If the length is multiplied by 3, the volume is multiplied by 3^3.

Hint: Use the table to explore what happens to the volume of a cube when you change the length of an edge.

Answer The volume of the full-size cedar chest will be 262,144 cubic inches.

Determine How did the pattern help you find the answer even though you did not know the dimensions of the model?

Sample Answer: The scale is 1:4. Four times the change in one dimension of a cube means the volume increases by a factor of 4^3, or 64.

③ Luke uses tiles to make a set of larger and larger square frames. How many tiles will he need altogether for a set of 7 frames, Frame 1 through Frame 7?

Ask Yourself: How many tiles are added each time?

Frame Number	1	2	3	4	5	6	7
Number of Tiles	8	12	16	20	24	28	32

Hint: Use the pattern to complete the table.

Answer Luke will need a total of 140 tiles.

Interpret Alicia found that Luke needs 32 tiles. What error did she probably make?

Sample Answer: Alicia found the number of tiles needed for the 7th frame, instead of the number of tiles needed for all 7 frames.

Lesson 14 Strategy Focus: Look for a Pattern   123

---

④ Ms. Keagan sells tool sheds. She is making a 1:10 scale model of her most popular shed. The total area of the four walls of the actual shed is 120 square feet. Ms. Keagan has 2 square feet of sample siding. Will that be enough to cover the walls of the model? How do you know?

Hint: Think of a wall as the face of a cube.

Ask Yourself: How can I find the relationship between a change in length and a change in area?

Length of 1 Face (ft)	1	2	3	...	10
Area of 1 Face (ft²)	1 × 1	2 × 2	3 × 3	...	10 × 10

Answer Yes, because the model will have a wall area of 1.2 square feet. 120 ÷ 100 = 1.2

Apply If Ms. Keagan used the scale 1:12, how would you find the wall area of the model?

Sample Answer: I'd use the pattern that the change in area is the square of the change in length. So the ratio of the wall areas would be 1:144.

⑤ Tucker is making a set of ten puzzle pieces. How many blocks will it take to make a set of 10 pieces, Piece 1 through Piece 10?

Piece 1   Piece 2   Piece 3

Ask Yourself: How does the number of blocks change from one piece to the next?

Piece	1	2	3	4	5	6	7	8	9	10
Number of Blocks	7	11	15	19	23	27	31	35	39	43

Hint: Use a pattern to help you find the sum of the number of blocks in each puzzle piece.

Answer It will take 250 blocks to make one 10-piece set.

Sequence What shortcut could you use when adding the numbers in the bottom row of the table to find the number of blocks in 10 puzzle pieces?

Sample Answer: I could find pairs of numbers with a sum of 50. Five of those pairs is 250 blocks.

124   Unit 4   Using Geometry

Practice

Solve the problems. Show your work.

6 Dana is building a set of models as shown. How many blocks will she need for the ninth model in the set?

Model 1   Model 2   Model 3

Some students will use the lesson strategy; however, other strategies may be used. Accept all reasonable work leading to the correct answer.

Answer She will need 100 blocks for the ninth model.

Discover Describe another pattern where the models would have the same number of blocks as each model in the set above.

Sample Answer: A set of squares starting with a 2 × 2 square, then a 3 × 3 square, then a 4 × 4 square, and so on will have the same number of blocks.

7 Kirk's class is creating a 1:15 scale model of an auditorium. If the real auditorium has a perimeter of 290 feet and a floor area of 4,500 square feet, what will be the perimeter and floor area of the model auditorium?

Some students will use the lesson strategy; however, other strategies may be used. Accept all reasonable work leading to the correct answer.

Answer The perimeter will be $19\frac{1}{3}$ feet, or 19 feet 4 inches. The floor area will be 20 square feet.

Relate Use the scale in this problem to explain how multiplying the dimensions of a rectangle by a factor relates to the changes in the perimeter and area of that rectangle.

Sample Answer: If you multiply the dimensions of a rectangle by 15, the perimeter is also multiplied by 15, and the area is multiplied by 15^2.

Create Make up a problem about a model made using a 1:10 scale that involves volume. Write the problem so that it can be solved by using the strategy *Look for a Pattern*. Solve your problem.

See teacher notes.

Lesson 14   Strategy Focus: Look for a Pattern   125

Create

In this lesson, students are asked to create their own problem about a model with a given scale that involves volume. If students are struggling, suggest they look back at Problem 2 for ideas.

Accept student responses that give a scale of 1:10 and the volume of an actual object or the volume of a model. Students should be able to use the strategy *Look for a Pattern* to arrive at a correct solution.

Independent Practice
Practice

Students should be encouraged to choose any strategy to solve Problems 6 and 7, though many may prefer to use *Look for a Pattern*.

6 Some students may find the next number of blocks by adding the next consecutive odd number to the previous number of blocks, while other students may write the next perfect square.

Sample Work

Model	1	2	3	4	5	6	7	8	9
Blocks	4	9	16	25	36	49	64	81	100

HOTS Discover Responses should show that students recognize that the pattern shows perfect squares and that they are able to create a pattern using those numbers of blocks.

7 Make sure students understand that they do not have to find the dimensions of the scale model or the actual auditorium to solve the problem.

Sample Work

Side Length	1	2	3	4	5
Perimeter	4	8	12	16	20
Area	1	4	9	16	25

290 ft ÷ 15 = $19\frac{1}{3}$ ft

4,500 ft² ÷ 15² = 20 ft²

HOTS Relate Students' explanations should indicate that if they multiply the dimensions of a rectangle by 15, the perimeter is multiplied by the same factor and the area is multiplied by the square of that factor.

Lesson 14   81

Lesson 15 Strategy Focus: Draw a Diagram

Lesson Overview

Lesson Materials: ruler, calculator

Skills to Know	Outcome	Math Vocabulary	eResources www.optionspublishing.com
• Identify and plot points on a coordinate grid • Find the center of dilation	Students will recognize that drawing a diagram is an efficient way to solve some problems about dilations.	dilation, enlarged, scale factor	• Interactive Whiteboard Transparency 15 • Homework, Unit 4 Lesson 15 • Know-Find-Use Table • Coordinate Grid 2 • Problem-Solving Checklist, also available in the student worktext, page 7

Modeled Instruction

Learn

Probe students' understanding of the problem's context by asking questions similar to the following.

- *How will the photo of the kite and the enlargement be the same? Different?*
- *What does it mean for something to be distorted?*

As students read the problem again, ask questions to help them focus on the details needed to solve the problem.

- *Where will you find information about the photo of the kite?*
- *Where will you find information about the enlargement of the photo?*

Use a Graphic Organizer You may wish to use this Think Aloud to demonstrate how to use a graphic organizer to identify information needed to solve the problem and plan how to use that information.

I know this problem is about a photo and its enlargement. Some information is given in the problem. Some of that information, and more, is given in the diagram.

I will read the problem again, filling in the Know-Find-Use Table to help me organize the details and see how I might use them to solve the problem. In the Know column, I can write △A'B'C' is an enlargement of △ABC. *I can also write the locations of each point that I know:* A(4, 5), A'(2, 6), B(8, 5), B'(10, 6), and C(6, 3). *In the Find column, I can write* The location of C'.

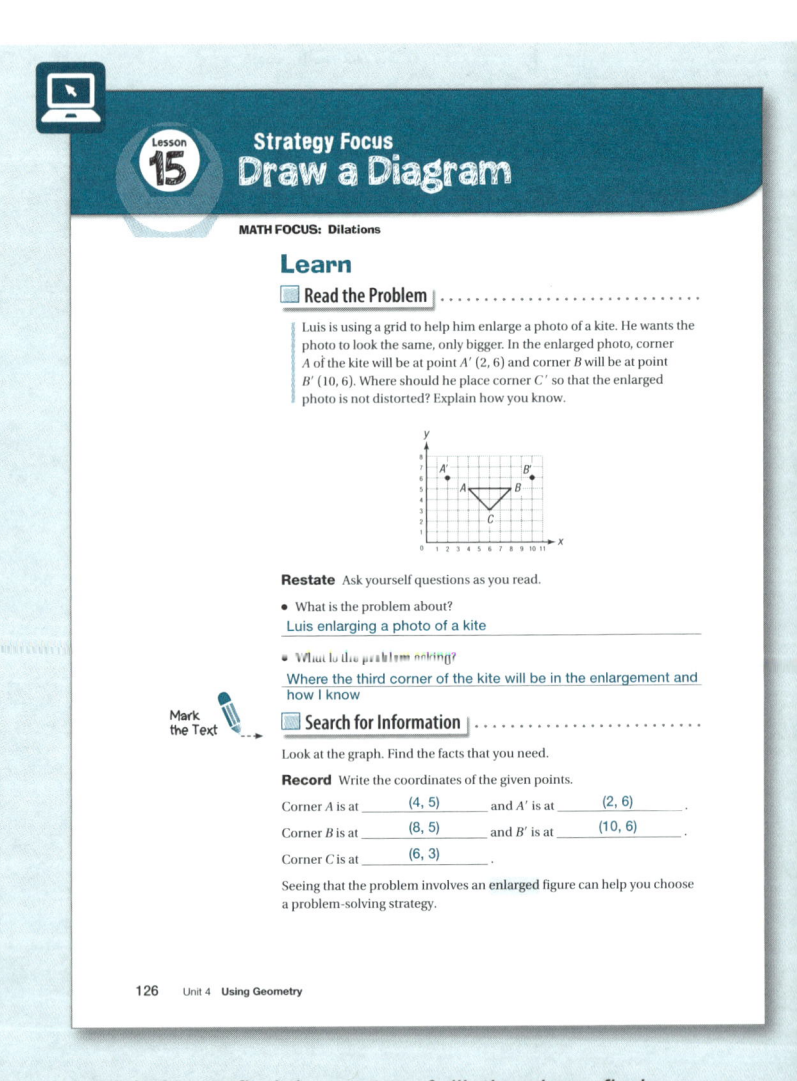

I think if I can find the center of dilation, I can find C'. In the Use column, I can write The locations of A and A' and B and B' to find the center of dilation.

Lesson 15 Strategy Focus: Draw a Diagram

MATH FOCUS: Dilations

Learn

Read the Problem

Luis is using a grid to help him enlarge a photo of a kite. He wants the photo to look the same, only bigger. In the enlarged photo, corner A of the kite will be at point A' (2, 6) and corner B will be at point B' (10, 6). Where should he place corner C' so that the enlarged photo is not distorted? Explain how you know.

Restate Ask yourself questions as you read.

- What is the problem about?
  Luis enlarging a photo of a kite

- What is the problem asking?
  Where the third corner of the kite will be in the enlargement and how I know

Mark the Text

Search for Information

Look at the graph. Find the facts that you need.

Record Write the coordinates of the given points.

Corner A is at _(4, 5)_ and A' is at _(2, 6)_.
Corner B is at _(8, 5)_ and B' is at _(10, 6)_.
Corner C is at _(6, 3)_.

Seeing that the problem involves an enlarged figure can help you choose a problem-solving strategy.

126 Unit 4 Using Geometry

---

82 Unit 4

Decide What to Do

You know the location of the original figure on the graph and you know two of the points of the enlargement.

Ask How can I find where to place corner C' of the enlarged kite?

- I can use the strategy *Draw a Diagram* to show where the points are and how they move.
- I can find the center of the dilation and determine how the figure changes when it enlarges.

Use Your Ideas

Step 1 Use the graph to make a diagram of the enlargement. Draw lines AA' and BB' to show how A and B move.

Step 2 Find the center of dilation.

The center of dilation is at __(6, 4)__. Then plot that point.

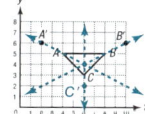

The center of dilation is where AA' and BB' intersect.

Step 3 Find the scale factor.

A' is __2__ times as far from the center of dilation as A.

B' is __2__ times as far from the center of dilation as B.

So the scale factor of the dilation is __2__.

Step 4 Plot C'. Draw a line through C and the center of dilation.

C' should be __2__ times as far from the center of dilation as C.

Corner C' of the enlarged photo will be at __(6, 2)__.

Review Your Work

Draw $\triangle A'B'C'$ on the diagram. It should be similar to $\triangle ABC$.

Explain How did drawing a diagram help you solve the problem?

Sample Answer: The diagram helped me find the center of dilation so I could find the scale factor and the location of C'.

127

Try

Solve the problem.

1. Mr. Palmer is planning the movement of dancers on a stage. The dancers will begin by forming $\triangle ABC$ and then move to form the larger $\triangle A'B'C'$. Dancer A will move to A', dancer B will move to B', and dancer C will move to C'. Where should dancer B move so $\triangle A'B'C'$ is similar to $\triangle ABC$?

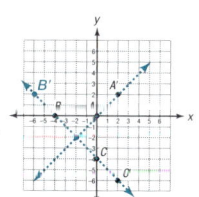

Mark the Text

Read the Problem and Search for Information

Relate the information in the problem to the graph.

Decide What to Do and Use Your Ideas

You can use the strategy *Draw a Diagram*. Use the graph to make a diagram.

Ask Yourself: Point A' is how many times as far from the center of dilation as point A?

Step 1 Draw lines through each original point and its corresponding point in the second formation.

Step 2 The center of dilation is __(−2, −2)__. Plot that point.

Step 3 Find the scale factor.

A' is __2__ times as far from the center of dilation as A.

Step 4 Plot B' by using the center of dilation and the scale factor.

Dancer B should move to __(−6, 2)__ so the two formations are similar.

Review Your Work

Draw $\triangle ABC$ and $\triangle A'B'C'$ to see if the two triangles are similar.

Describe If you wanted to find the area of the larger triangle, why would drawing a diagram still be a useful strategy?

Sample Answer: A diagram would show the size of the large triangle. Then I could use the base and height to calculate the area.

128 Unit 4 Using Geometry

Modeled Instruction (continued)

Help students recognize how the details of the problem can be used to choose a problem-solving strategy.

- *Is $\triangle A'B'C'$ a dilation of $\triangle ABC$? How do you know?*
- *How can knowing the center of dilation help you determine how the figure changes when it is enlarged?*

Ask questions that guide students to consider each step in the solution process.

- *Why is the ordered pair (6, 4) the location of the center of dilation?*
- *Why does C' need to be 2 times as far from the center of dilation as C?*

Ask students how they can check if the dilation is similar to the original figure.

HOTS Explain Students' explanations should mention the importance of the diagram in finding the center of dilation and the scale factor.

Guided Practice

Try

1. Prompt students to consider how this problem is similar to the previous problem.

Discuss the context of the problem, as well as the information needed to solve it.

- *What words or phrases tell you this problem involves a dilation?*

Help students through the steps of finding the solution.

- *How did you find the scale factor?*
- *How can you use the scale factor to find the coordinates of B'?*

Have students explain how they can tell if the triangles are similar by drawing them.

HOTS Describe Students' responses should indicate that drawings can help them visualize the situation and ensure that they use the correct lengths for the sides of the triangle.

Lesson 15 83

Scaffolded Practice
Apply

② Make sure students understand that in this problem, the dilation is a reduction instead of an enlargement.

- How is this problem different from the problems you have already solved in this lesson?
- Will these differences change the steps you will use to solve the problem? Why or why not?

HOTS Illustrate Students' explanations should demonstrate an understanding of the importance of making accurate drawings to find an exact answer.

③ Make sure students see how the points A, B, C, and D relate to points A', B', C', and D'.

- Why do you need to plot A, B, C, and D and their images?

HOTS Determine Responses could indicate that students find it easier to see the dilation if they can relate it to similar figures.

④ Make sure students recognize that the drum major's position, B, does not change.

- How can you find the center of dilation when a point is in the same location as its image?

HOTS Identify Responses should show that students find that *Draw a Diagram* is a useful strategy when they have to work with points on a coordinate grid.

⑤ Make sure students understand that they only have to look at the individual points given to make the new drawing.

- Why can you use the locations of points A and B and A' and B' to find the new location of a point on the triangle?

HOTS Decide Students' explanations should mention that the ratios of the lengths of corresponding parts of similar figures are equal so the width will also increase.

Apply
Solve the problems.

② Courtney is making a collage of rectangles. Her first rectangle ABCD is shown. She wants rectangle A'B'C'D' to be similar, but smaller. A' will be at (8, 12) and B' at (8, 10). Where will she plot C'? Explain how you know.

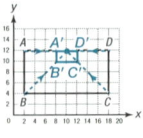

Draw lines AA' and BB'.

The center of dilation is ____(10, 12)____.

The scale factor of this problem is __$\frac{1}{4}$__.

Answer She will plot C' at (12, 10). Each point moves to $\frac{1}{4}$ its distance from the center of dilation.

Ask Yourself Are the corners moving toward the center of dilation or away from it?

Hint Use the center of dilation to determine how C will move.

Illustrate Why is it important to draw the lines accurately?

Sample Answer: If the lines are not drawn accurately, you could end up with an incorrect center of dilation or an incorrect scale factor.

③ Mr. Tan is planning a light show in which a geometric figure appears to expand to form a larger, similar figure. First, lights A, B, C, and D will be on. Then they will turn off and lights A', B', C', and D' will turn on. The table shows the placement of the lights. What should be the coordinates of B'?

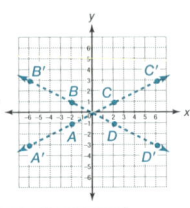

A (−2, −1)	B (−2, 1)	C (2, 1)	D (2, −1)
A' (−6, −3)	B' (−6, 3)	C' (6, 3)	D' (6, −3)

Draw lines AA' and DD'.

Answer The coordinates of B' should be (−6, 3).

Ask Yourself How can I use the diagram to find the scale factor?

Hint Plot the points on the coordinate grid.

Determine How would drawing quadrilaterals ABCD and A'B'C'D' help you to solve this problem?

Sample Answer: Drawing the quadrilaterals would help me check that the figures are similar, as they should be in a dilation.

Lesson 15 **Draw a Diagram** 129

---

④ A marching band starts in the formation shown. The tuba player is at A, the drum major is at B, and the drummer is at C. In the second formation, the drum major will be at (0, 5) and the tuba player at (0, 25). Where should the drummer be so that the second formation is similar to the first? Explain how you know.

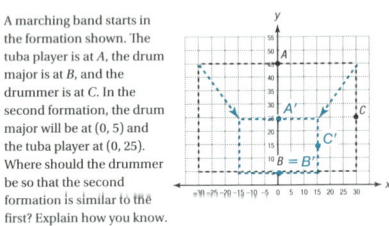

The scale factor of the dilation is $\frac{1}{2}$.

The center of the dilation is ____(0, 5)____.

Answer The drummer should be at (15, 15). She should be half of the distance that she was from the drum major.

Ask Yourself How long are the sides of the new formation?

Hint Draw a rectangle showing the new formation.

Identify What is it about the problem that indicates *Draw a Diagram* will be a useful strategy?

Sample Answer: The problem involves the old and new location of points.

⑤ Lexi is enlarging a photo of a building. Point A will move to (1, 3) and point B to (7, 3). Where will the top of the spire, Point T, move to? Explain how you know.

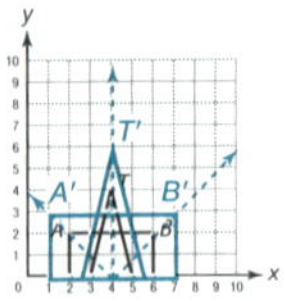

Plot where points A and B will move to.

The center of dilation is ____(4, 0)____.

Answer The top of the spire will be at (4, 6). I know because the scale factor is $\frac{3}{2}$.

Ask Yourself Where is the center of dilation?

Hint Draw lines that show the dilation.

Decide Does the base of the spire stay the same width? Explain.

Sample Answer: No, because all points within the photo will move by a scale factor of 3:2 centered on (4, 0).

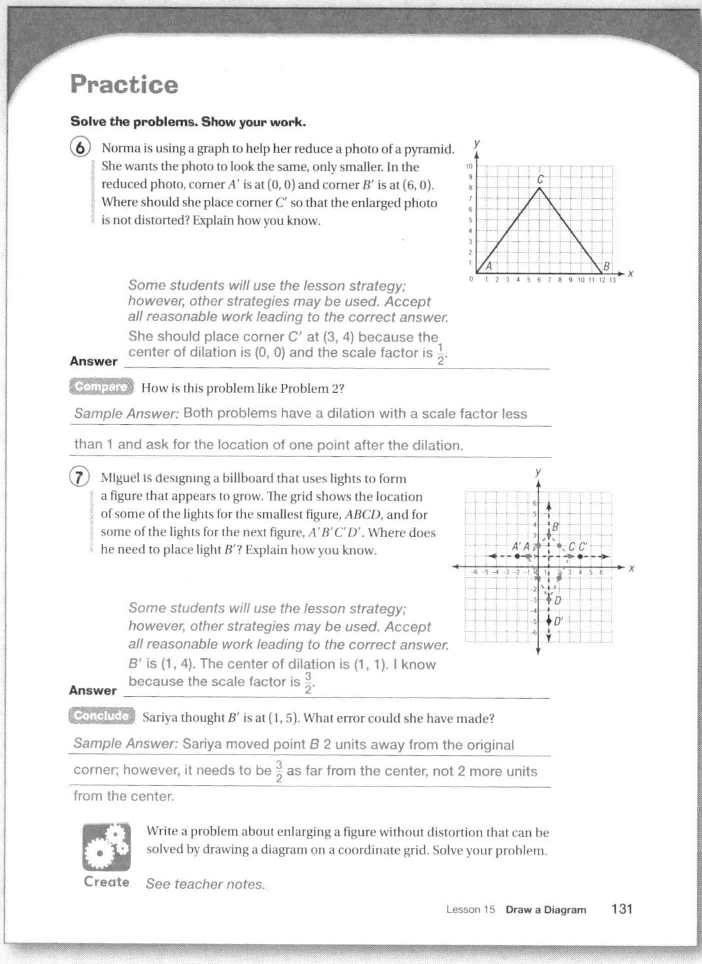

Practice

Solve the problems. Show your work.

6 Norma is using a graph to help her reduce a photo of a pyramid. She wants the photo to look the same, only smaller. In the reduced photo, corner A' is at $(0, 0)$ and corner B' is at $(6, 0)$. Where should she place corner C' so that the enlarged photo is not distorted? Explain how you know.

Answer
Some students will use the lesson strategy; however, other strategies may be used. Accept all reasonable work leading to the correct answer.
She should place corner C' at $(3, 4)$ because the center of dilation is $(0, 0)$ and the scale factor is $\frac{1}{2}$.

Compare How is this problem like Problem 2?
Sample Answer: Both problems have a dilation with a scale factor less than 1 and ask for the location of one point after the dilation.

7 Miguel is designing a billboard that uses lights to form a figure that appears to grow. The grid shows the location of some of the lights for the smallest figure, $ABCD$, and for some of the lights for the next figure, $A'B'C'D'$. Where does he need to place light B'? Explain how you know.

Answer
Some students will use the lesson strategy; however, other strategies may be used. Accept all reasonable work leading to the correct answer.
B' is $(1, 4)$. The center of dilation is $(1, 1)$. I know because the scale factor is $\frac{3}{2}$.

Conclude Sariya thought B' was at $(1, 5)$. What error could she have made?
Sample Answer: Sariya moved point B 2 units away from the original corner; however, it needs to be $\frac{3}{2}$ as far from the center, not 2 more units from the center.

Create Write a problem about enlarging a figure without distortion that can be solved by drawing a diagram on a coordinate grid. Solve your problem.
See teacher notes.

Lesson 15 Draw a Diagram 131

Create

In this lesson, students must write a problem about a dilation. If students are struggling, suggest they draw a figure and a dilation, then write a problem that does not give one point in the image.

Accept student responses that include a problem about an enlargement that is correctly solved using coordinates. Information may be given in the problem, a diagram, or both. Enough information must be given to locate a center of dilation.

Independent Practice
Practice

Students should be encouraged to choose any strategy to solve Problems 6 and 7, though many may prefer to use *Draw a Diagram*.

6 Some students may choose to find the scale factor and the center of dilation without a diagram using the fact that $A = A'$ and the scale factor is $\frac{1}{2}$.

Sample Work

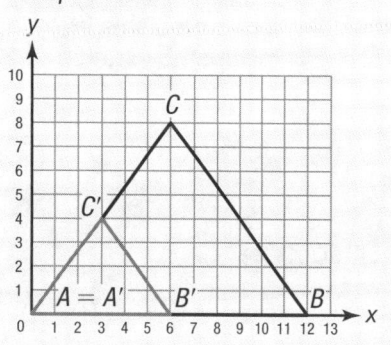

Compare Responses should show that students recognize that both problems involve a reduction and ask for the location of the image of a point.

7 Students should recognize that the diagram already shows the center of dilation.

Sample Work

A' is 1.5 times as far from the center of dilation as A.

So B' needs to be 1.5 times as far from the center of dilation as B.

$1.5 \times 2 = 3$, so B' must be 3 units above the center of dilation.

Conclude Responses should show that students can recognize a common error that is made when solving problems involving dilations.

Lesson 15   85

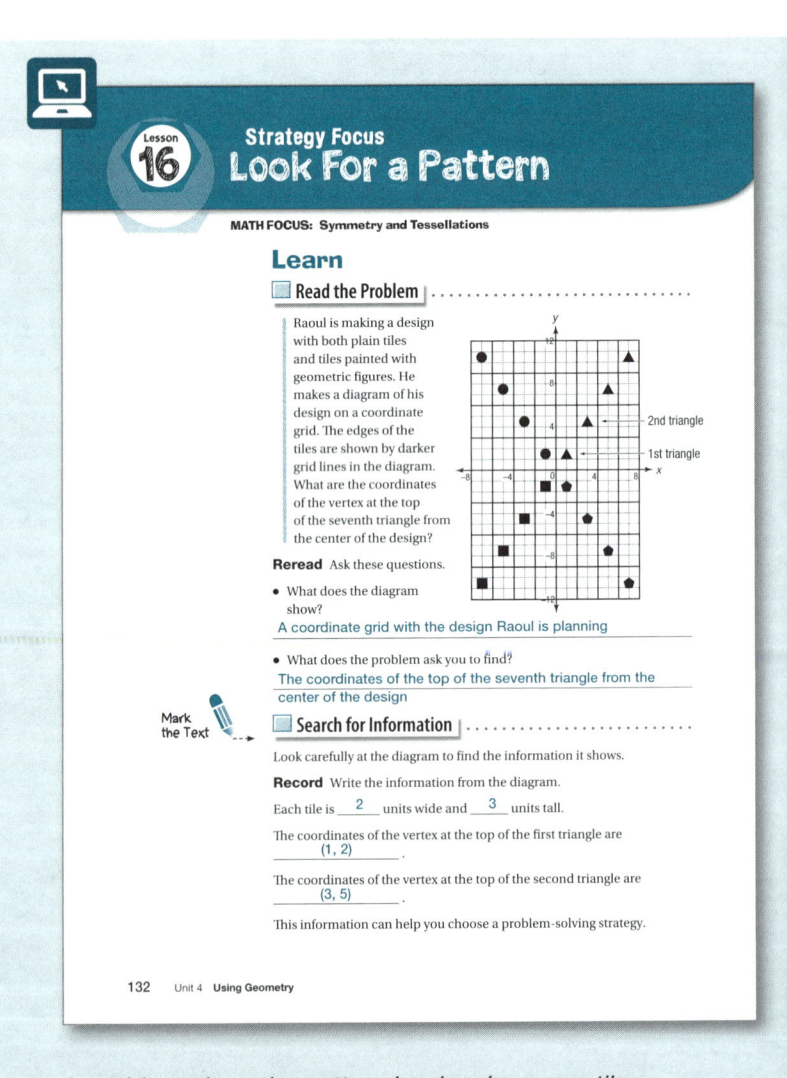

Lesson 16

Strategy Focus
Look for a Pattern

Lesson Overview

Lesson Materials: ruler, calculator

Skills to Know	Outcome	Math Vocabulary	eResources www.optionspublishing.com
• Identify and plot points on a coordinate grid • Recognize line symmetry in figures	Students will recognize that looking for a pattern is an efficient way to solve problems involving geometric patterns.	symmetry, tessellation	• Interactive Whiteboard Transparency 16 • Homework, Unit 4 Lesson 16 • Coordinate Grid 2 • Problem-Solving Checklist, also available in the student worktext, page 7

Modeled Instruction

Learn

Ask questions about the problem's context to clarify students' comprehension of what the problem is about.

- Does Raoul's design follow a pattern? What is the pattern?
- Is all of Raoul's design shown in the diagram? How do you know?

As students read the problem again, pose questions to help them recognize important phrases and facts.

- How can you find the coordinates of the vertex at the top of the first triangle?

Reread You may wish to use this Think Aloud to show how to create a plan for solving the problem.

I know this problem is about a tile design. Part of it is shown in the diagram. I need to find out information about a triangle that would be in the design if the design was larger.

I'm going to read the problem again to be sure I know what information I need to find. Then I will study the diagram closely.

I see that I need to find the coordinates of the top of the seventh triangle. I see in the diagram that the tiles with triangles are arranged in a pattern that makes a diagonal. The triangles are in the same place on each of these tiles, so the triangles make a pattern.

I could continue the pattern by drawing more tiles until I have 7 triangles. But, since the design is drawn on a coordinate grid, I wonder if I can use numbers to describe this pattern.

86   Unit 4

Decide What to Do

You know the design of the tiles. You know the length and width of each tile, and the coordinates of the tops of the first four triangles.

Ask How can I find the coordinates of the top of the seventh triangle?

- I can organize the coordinates in a table, then use the strategy *Look for a Pattern*.

> When something keeps changing predictably, you can look for a pattern.

Use Your Ideas

Step 1 Write the coordinates for the vertices at the tops of the first four triangles. Fill in the table.

Number of Triangle	1st	2nd	3rd	4th	...	7th
x-coordinate	1	3	5	7		13
y-coordinate	2	5	8	11		20

Step 2 Look for a pattern that describes the coordinates of the vertices at the tops of the triangles.

The x-coordinate increases by __2__ between two triangles.

The x-coordinate is 2 times the number of the triangle minus 1.

So the x-coordinate of the vertex at the top of the seventh triangle is $2 \times$ __7__ $- 1$, or __13__.

The y-coordinate increases by __3__ between two triangles.

The y-coordinate is 3 times the number of the triangle minus 1.

So the y-coordinate of the vertex at the top of the seventh triangle is $3 \times$ __7__ $- 1$, or __20__.

The coordinates of the top of the seventh triangle are __(13, 20)__.

Review Your Work

Sketch more tiles to verify the pattern you found.

Explain How did making a table help you see the pattern?

Sample Answer: Making a table helped me see that the coordinates change in a predictable way.

Modeled Instruction (continued)

Help students make a connection between the facts and a strategy they can use to solve the problem.

- *What pattern do you see in the coordinates of the vertex at the top of the first four triangles?*

Ask questions that encourage students to think critically about the steps in the solution process.

- *How did you know which information to use to fill in the table?*

Emphasize to students the importance of checking that they have found the correct pattern.

- *How can drawing the fifth tile with a triangle help you check that your pattern is correct?*

HOTS Explain Students' explanations should point out that a table provides an organized way of recording numbers so that patterns can be seen more easily.

Try

Solve the problem.

① Katelyn is making a patchwork quilt. The quilt will have 18 rings of pentagons and triangles around a pentagon in the center. How many pentagons will Katelyn need for the 18th ring?

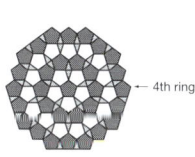

← 4th ring

Read the Problem and Search for Information

Identify what the problem asks you to find. Reread, then circle the important data in the problem.

Decide What to Do and Use Your Ideas

You can use the strategy *Look for a Pattern*.

Step 1 Find the number of pentagons in the first four rings.

Use **symmetry** to help you count the number of pentagons in each ring of the **tessellation**.

Ring Number	1	2	3	4	...	18
Number of Pentagons	5	10	15	20		90

Step 2 Identify and use the pattern.

Number of pentagons in ring = __5__ × the ring number.

The 5th ring would have 5 × __5__, or __25__, pentagons.

The 6th ring would have 5 × __6__, or __30__, pentagons.

The 18th ring would have 5 × __18__, or __90__, pentagons.

Katelyn will need __90__ pentagons for the 18th ring.

Review Your Work

Make sure you answered the question that was asked.

Describe What shortcut can you use to quickly add the numbers in the second row of the table to find the total number of pentagons?

Sample Answer: I can add 9 pairs of numbers that add up to 95, for example, 5 + 90 and 10 + 85. I can multiply 95 times 9 and then add 1 for the pentagon in the center.

> *Ask Yourself:* What is the relationship between the number of pentagons in a ring and the ring number?

Guided Practice

Try

① Prompt students to consider the relationship between the diagram given and the table they will complete.

Discuss the context of the problem as well as the information given in the diagram.

- *How are the rings distinguished from one another?*

Have students explain how they completed the table and used it to answer the question the problem asks.

- *How did the completed table help you solve the problem?*

Make sure students find the number of pentagons in the 18th ring, not the total number of pentagons.

HOTS Describe Students' responses should describe a shortcut for finding the number of pentagons needed for all 18 rings.

Scaffolded Practice
Apply

② Make sure students understand that the rows are numbered from the top down and that, starting at the second row, each row contains gray and white triangle tiles.

- How can you describe in words the way Sue is creating this design?

Predict Students' explanations should indicate that they recognize the pattern and can use it to find the number of tiles in any given row.

③ Help students solve the problem by relating it to a previous problem.

- How is this problem like Problem 1? How is it different?

Conclude Responses should indicate that students recognize that the table does not show the answer, and that more computation is necessary to solve the problem.

④ Prompt students to see the relationship between (17, 17) and the points marked in the diagram.

- Why will it be easier to look for a pattern instead of drawing more squares and points?

Analyze Responses should indicate that students can identify the pattern for each row of the table.

⑤ Make sure students understand what is meant by the phrase *the coordinates of the lower two vertices*.

- What are the coordinates of the lower two vertices of the triangle design made up of the 4 rows shown?

Identify Responses should show that students understand that when changes continue in the same way, it can be helpful to look for a pattern.

Apply
Solve the problems.

② Sue is creating this design using triangular tiles. If she continues until she has 20 rows of tiles, how many triangular tiles will Sue have in the 20th row?

Row Number	1	2	3	4	...	20
Number of Tiles	1	3	5	7		39

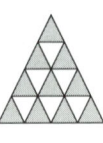

Ask Yourself How does the number of tiles increase between rows?

Hint Use words to describe the relationship between the number of tiles in a row and the row number.

Answer Sue will have 39 triangular tiles in the 20th row.

Predict Describe how can you use the pattern to find the number of tiles she will need for the 25th row.

Sample Answer: The number of tiles in a row is always 2 times the number of the row minus 1. So Sue will need 2(25) − 1, or 49 tiles for the 25th row.

③ Mr. Ross is tiling the floor of a circular lobby as shown. The first dark ring contains 6 tiles. There will be 8 dark tile rings. How many dark tiles will Mr. Ross need for all of the 8 dark tile rings?

Dark Tile Ring	1st	2nd	3rd	4th	...	8th
Tile	6	18	30	42		90

Hint You can use the symmetry of the design to count the tiles more easily.

Ask Yourself How does the number of tiles increase between the rings of dark tiles?

Answer Mr. Ross will need 384 dark tiles for the 8 dark tile rings.

Conclude Nicholas says that the answer to this problem is 90 tiles. What error could Nicholas have made?

Sample Answer: Nicholas found the number of tiles in the 8th dark ring, instead of the sum of the tiles in the 8 dark tile rings.

Lesson 16 **Look For a Pattern** 135

Ask Yourself Can I find a pattern to answer the question without drawing more rows in the diagram?

④ Chris is plotting this design on a coordinate grid. If he continues his pattern, will the point (17, 17) be inside a light or a dark square? How do you know?

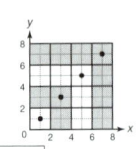

Light Square	(1, 1)	(5, 5)	(9, 9)	(13, 13)	(17, 17)
Dark Square	(3, 3)	(7, 7)	(11, 11)	(15, 15)	(19, 19)

Hint Identify the marked points and whether they are located in light or dark squares.

Answer The point (17, 17) will be inside a light square. I extended the pattern until I got to (17, 17).

Analyze How can you use the pattern in the table to find whether any point where both coordinates are equal and odd will fall in a light or a dark square?

Sample Answer: If the number used in the coordinate pair is 1 more than a multiple of 4, it will fall in a light square. If it is 1 less than a multiple of 4, it will fall in a dark square.

Ask Yourself How can I solve this without having to draw the rest of the 14 rows?

⑤ Dan is drawing this triangle design on a coordinate grid. He will keep adding rows of triangles under the 4 rows already drawn. If he continues the design to a total of 14 rows, what will the coordinates of the lower two vertices be?

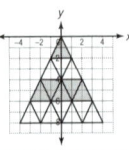

Hint Look at how the row number in each column is related to the x-coordinates and the y-coordinates in that column.

Row Number	1	2	3	4	...	14
Left Vertex	(−1, −2)	(−2, −4)	(−3, −6)	(−4, −8)		(−14, −28)
Right Vertex	(1, −2)	(2, −4)	(3, −6)	(4, −8)		(14, −28)

Answer The left vertex will be (−14, −28) and the right vertex will be (14, −28).

Identify How did you know that *Look for a Pattern* would be a useful strategy?

Sample Answer: The problem asked about where something would be if it kept changing in the same way.

Practice

Solve the problems. Show your work.

6 Phil is covering a patio with white and gray trapezoid-shaped bricks. If Phil continues the design in the same way, how many bricks will he need for the 5th white ring?

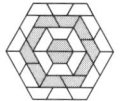

Some students will use the lesson strategy; however, other strategies may be used. Accept all reasonable work leading to the correct answer.

Answer Phil will need 38 bricks for the 5th white ring. Each ring has 4 more than the ring before.

Assess How can you check your work for this problem?

Sample Answer: I can check against a pattern of all of the rings. Each ring adds 4 more trapezoids, so the 5th white ring is 8 rings outside the 1st white ring and has 8 × 4 + 6 = 38 tiles.

7 Barbara is planning an animation of a honeycomb tessellation on a coordinate grid. There is a worker bee in the third ring of hexagons, as shown. The queen bee will fly through the centers of the hexagons in the twelfth ring. Where will the queen cross the x-axis?

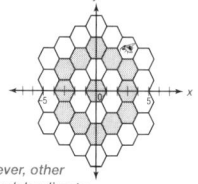

Some students will use the lesson strategy; however, other strategies may be used. Accept all reasonable work leading to the correct answer.

Answer The queen will cross the x-axis at $(-18, 0)$ and $(18, 0)$.

Compare How is this problem like Problem 4? How is it different?

Sample Answer: Both problems use a pattern in the coordinates of centers of pieces. The patterns in the two problems are different.

Create Write a problem about a symmetric design using squares that can be solved by finding a pattern. Write and solve your problem.

See teacher notes.

Lesson 16 Look For a Pattern 137

Create

In this lesson, students write a new problem similar to ones they solved. If students are struggling, suggest they look back at the problems in this lesson and make a square design with square tiles.

Accept student responses that include a problem about a design using squares that can be solved using a pattern. A pattern should be evident in the design and a correct solution should be provided. Solutions should be the result of using that pattern.

Independent Practice
Practice

Students should be encouraged to choose any strategy to solve Problems 6 and 7, though many may prefer to use *Look for a Pattern*.

6 Some students may solve this problem by drawing a diagram.

Sample Work

White Tile Ring	1st	2nd	...	5th
Number of Tiles	6	14	...	38

HOTS Assess Students' explanations should demonstrate an understanding of how to check values in a pattern.

7 Make sure students recognize that the position of the worker bee in the third ring indicates how the rings are numbered.

Sample Work

Ring Number	1	2	3	...	12
Center of Ring on Left Side	(−1.5, 0)	(−3, 0)	(−4.5, 0)		(−18, 0)
Center of Ring on Right Side	(1.5, 0)	(3, 0)	(4.5, 0)		(18, 0)

HOTS Compare Responses should show that students recognize likenesses and differences in similar problems.

Lesson 16 89

UNIT 4 Review

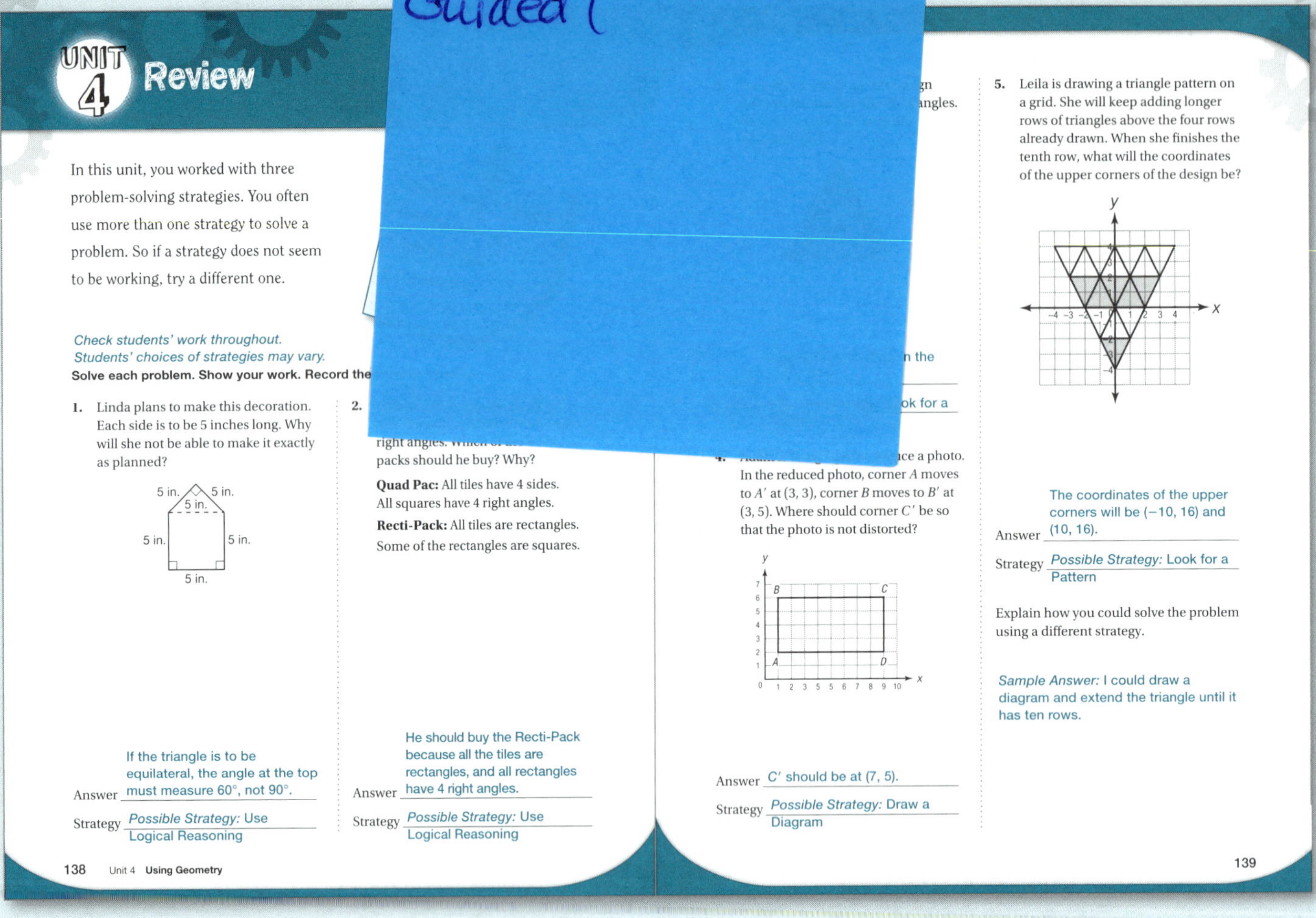

UNIT 4 Review

In this unit, you worked with three problem-solving strategies. You often use more than one strategy to solve a problem. So if a strategy does not seem to be working, try a different one.

Check students' work throughout.
Students' choices of strategies may vary.
Solve each problem. Show your work. Record the...

1. Linda plans to make this decoration. Each side is to be 5 inches long. Why will she not be able to make it exactly as planned?

   5 in. 5 in.
   5 in.
   5 in. 5 in.
   5 in.

   Answer: *If the triangle is to be equilateral, the angle at the top must measure 60°, not 90°.*
   Strategy: *Possible Strategy: Use Logical Reasoning*

2. ...
   Quad Pac: All tiles have 4 sides. All squares have 4 right angles.
   Recti-Pack: All tiles are rectangles. Some of the rectangles are squares.

   ...right angles. Which of ... packs should he buy? Why?

   Answer: *He should buy the Recti-Pack because all the tiles are rectangles, and all rectangles have 4 right angles.*
   Strategy: *Possible Strategy: Use Logical Reasoning*

4. ...ce a photo. In the reduced photo, corner A moves to A' at (3, 3), corner B moves to B' at (3, 5). Where should corner C' be so that the photo is not distorted?

   Answer: *C' should be at (7, 5).*
   Strategy: *Possible Strategy: Draw a Diagram*

5. Leila is drawing a triangle pattern on a grid. She will keep adding longer rows of triangles above the four rows already drawn. When she finishes the tenth row, what will the coordinates of the upper corners of the design be?

   Answer: *The coordinates of the upper corners will be (−10, 16) and (10, 16).*
   Strategy: *Possible Strategy: Look for a Pattern*

   Explain how you could solve the problem using a different strategy.

   Sample Answer: I could draw a diagram and extend the triangle until it has ten rows.

138 Unit 4 Using Geometry

139

Support for Assessment

The problems on pages 138–141 reflect strategies and mathematics students used in the unit.

Although this unit focuses on three problem-solving strategies, students may use more than one strategy to solve the problems or use strategies different from the focus strategies. Provide additional support for those students who need it.

Use Logical Reasoning For Problem 1, ask students what they know about the angles of an equilateral triangle. For Problems 2 and 10, ask students to identify the statements that tell about the shapes in the packs and what shapes could be in each pack.

Look for a Pattern For Problems 3, 5, 6, 8, and 9, ask students to describe the patterns they find. For Problems 3 and 6, they can extend the patterns as needed. For Problems 5 and 8, they should apply the pattern to the row or ring number. For Problem 8, ask students to identify the final step in answering the question.

Draw a Diagram For Problems 4 and 7, ask questions such as, *What do you need to find before you can find the location of the missing point in the image? How will a diagram help you do this?*

You may wish to use the *Review* to assess student progress or as a comprehensive review of the unit.

Promoting 21st Century Skills

Write About It
Communication

When students describe the process of solving a problem, they have the opportunity to review their decisions on how to approach the problem. Students' responses should clearly state each step they used to reach their solution.

Team Project: Tessellation Nation
Collaboration: 3–4 students

Remind students that the goal of the project is to make a group decision, work together to carry out their plan and make a group presentation of the work.

Monitor groups as they work to check that each member is participating in all phases of the project.

Ask questions that help students summarize their thinking. *How did you organize your work? Does your figure change in the same way from one design to the next? Explain.*

Extend the Learning
Media Literacy

If you have Internet access, navigate to sites where students can access computer drawing programs. They may find them helpful in creating diagrams of their designs to use in the group presentations.

🔍 Search tessellations

Solve each problem. Show your work. Record the strategy you use.

6. Albert is making larger and larger designs that all have the same shape. How many tiles will he need to make the 5th design in the series?

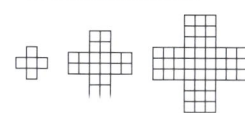

   Answer: *He will need 125 tiles.*
   Strategy: *Possible Strategy: Look for a Pattern*

7. Reba is enlarging square ABCD. She is using a scale factor of 2:5. The center of dilation is (2, 1). In the enlarged square, what will the coordinates of corner D' be?

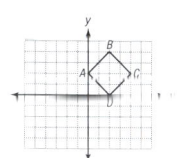

   Answer: *Corner D' will be at (2, −4).*
   Strategy: *Possible Strategy: Draw a Diagram*

8. Dylan is making this design. How many triangles will he need in all to extend the design to the third white ring?

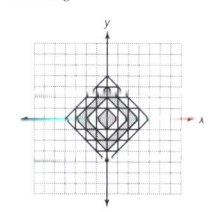

   Answer: *He will need 144 triangles in all.*
   Strategy: *Possible Strategy: Look for a Pattern*

   Explain how you found the total number of triangles.

   Sample Answer: I noticed that each quadrant of the graph was identical, and the number of triangles in each ring grew in the pattern 1, 3, 5, 7, … So the total number of triangles is 4 × (1 + 3 + 5 + 7 + 9 + 11), which is 144.

9. Mr. Lyon wants to make a 1:12 scale model of a log cabin. If the length of the side of the actual cabin is 18 feet and the length of the front is 12 feet, what will the area of the floor of the model be?

   Answer: *The area will be 1.5 square feet.*
   Strategy: *Possible Strategy: Look for a Pattern*

10. Cathy is selecting tiles for a project. Which pack of tiles would have the greater variety of shapes?

    Right-Angle Pack: All tiles are quadrilaterals. Every angle is a right angle.

    Equi-Pack: Every tile is equilateral. Some tiles have more sides than other tiles.

    Answer: *The Equi-Pack would have the greater variety of shapes.*
    Strategy: *Possible Strategy: Use Logical Reasoning*

Write About It
Look back at Problem 3. Describe how you solved the problem.

Sample Answer: I saw that each row had 2 more triangles than the row above it. When I made a table to show that pattern, I noticed that the number of triangles in each row was 1 less than twice the row number. So the row with 25 triangles would be the 13th row.

Team Project: Tessellation Nation

Your team will plan a growing design. Your plan will describe and show a series of figures that can be constructed from congruent copies of one tile shape. As a group, choose a tile shape from the list at the right. You will use copies of that shape according to a pattern that adds a growing number of the shape to each new design. Draw the first five figures of the series. Your design can have gaps, but no overlaps.

Plan — Read the problem and understand the options.
Decide — Choose a shape to use and what growing design to create from it
Justify — Explain your group's pattern. Use tables, lists, or diagrams.
Present — As a group, share the plan of your design. Discuss your reasoning.

Tile Shapes
- Equilateral Triangle
- Rectangle
- Square
- Rhombus
- Trapezoid
- Regular Pentagon
- Regular Hexagon
- Regular Octagon

Unit 4   91

UNIT 5: Problem Solving Using Measurement

CCSS 8.G Geometry

Unit Overview

Lesson	Problem-Solving Strategy	Math Focus
17	Draw a Diagram	Perimeter and Circumference
18	Solve a Simpler Problem	Area
19	Guess, Check, and Revise	Surface Area and Volume
20	Write an Equation	Similarity

Promoting Critical Thinking

Higher order thinking questions occur throughout the unit and are identified by this icon: HOTS. These questions progress through the cognitive processes of remembering, understanding, applying, analyzing, evaluating, and creating to engage students at all levels of critical thinking.

UNIT 5: Problem Solving Using Measurement

Unit Theme: Monuments and Landmarks

You probably know about the Statue of Liberty and the Lincoln Memorial. You may have never heard of the Sipapu Bridge or Kanton Island. In this unit, you will see how math is relevant to all kinds of structures. These structures can be natural or artificial, ancient or new.

Math to Know

In this unit, you will apply these math skills:
- Determine perimeter, circumference, and area of two-dimensional shapes
- Find surface area and volume of three-dimensional shapes
- Use similar figures to find measures

Problem-Solving Strategies
- Draw a Diagram
- Solve a Simpler Problem
- Guess, Check, and Revise
- Write an Equation

Link to the Theme

Write another paragraph about Max's drawing of Stonehenge. Use words that would help a person visualize what you see in the picture.

Max is giving an oral presentation about the ancient monument in England called Stonehenge. He accidentally left his drawing of it at home. So he describes what it looks like to the class.

Students' paragraphs will vary, but should include some words that describe the picture.

Use Math Language

Review Vocabulary

The list below shows vocabulary terms in this unit. Knowing the meaning of these terms will help you understand the problems.

base   diameter   perimeter   proportion
circumference   equivalent ratios   prism   pyramid

Vocabulary Activity — Word Pairs

Related math terms that are taught together may mean different things. Use terms from the above list to complete the following sentences.

1. **Distance Around**
   a. Jamie measures the distance around a pond to find its _____circumference_____.
   b. If Darrell walks along the edge of a rectangular garden plot to measure it, he would walk along the garden's _____perimeter_____.

2. **Solid Figures**
   a. A _____prism_____ is a solid figure with two parallel faces that are congruent polygons. The remaining faces are rectangles.
   b. A _____pyramid_____ is a solid figure in which one face is a polygon. The remaining faces are triangles that meet at a point.

Graphic Organizer — Word Map

Complete the graphic organizer.
- Write a definition of *diameter*.
- Use the term in a sentence.
- Show an example and a non-example of the term.

Sample Answers:

Definition	Sentence
a line segment that passes through the center of a circle, and whose end-points are two points on the circle	The diameter of the Ferris wheel measures 250 feet.
Example	Non-Example

Link to the Theme Monuments and Landmarks

Ask students to read the direction line and story starter. If students are having trouble getting started, ask questions such as, *What monuments have you seen? What did they look like? How would you describe them?*

Unit 5 Differentiated Instruction

Extra Support

Some students may need reinforcement of geometry concepts and proportions.

Pythagorean Theorem Provide students with a right triangle having sides of 3 cm, 4 cm, and 5 cm. Have students cut a 3-cm, a 4-cm, and a 5-cm square from a sheet of centimeter grid paper (Grid Paper – 1 cm). Have them place the squares along the sides of the triangles so that the side of a square matches the side of the triangle. Then have students find the area of each square and determine how the areas are related. Repeat for a triangle with sides of 5 cm, 12 cm, and 13 cm. Relate student findings about the areas of the squares to the Pythagorean Theorem.

Proportions Have students draw two sets of shapes: one set with 4 circles and 8 squares and one with 6 circles and 12 squares. Guide students to see that the ratio of circles to squares in each set is the same. Write these possible proportions: $\frac{4}{8} = \frac{6}{12}$, $\frac{4}{6} = \frac{8}{12}$, $\frac{4}{12} = \frac{6}{8}$, and $\frac{4}{8} = \frac{12}{6}$. Have students use cross multiplication to find which ratios actually form proportions. Compare the order of terms in the ratios in those equations that are proportions and in those that are not. Challenge students to write another proportion using the circles and squares, such as $\frac{8}{4} = \frac{12}{6}$.

Challenge Early Finishers

Early finishers may enjoy the challenge of solving open problems with multiple solutions. You can make any problem open by removing the question and asking students what they can figure out. For example:

> Jon is planning the design for a new skateboarding park. The park is a 300-meter by 180-meter rectangle with a semicircular plaza on one of the short sides. The radius of the plaza is 90 meters.

Ask students to consider how the information in the problem is related and what questions they can ask about the situation. Then have students write a question for the problem and answer it.

English Language Learners

Vocabulary

Circle, Semicircle, Crescent On the board, draw a circle, prompt students to tell you it is a circle, then label the drawing *Circle*. Next, draw a second circle, draw a line down its middle, and erase one curved side so that a semicircle remains. Explain that this is a semicircle, or half circle, and that *semi-* is a prefix that means *half*. Label the drawing *Semicircle* and have students say the word. Then draw a third circle and transform it into a crescent by drawing a curved line inside the circle and erasing the extraneous part of the circle. Label the drawing *Crescent*, and have students say the word. Finally, have students draw and label each of the three shapes at their own desks.

Build Background

Visualizing Structures To help students visualize the complex shapes in this unit, show students pictures from books or the Internet of the real structures described in some of the problems. As you show structures such as the White House or the Tower of Pisa, have students use math language involving shape, size, height, and so on, to describe the structures. Also point out that some of the problems ask you to visualize an object from an aerial view. Be sure students know this term means viewing a structure from above.

Language Development

Compound Modifiers Write the following phrase from the unit on the board: *crescent-shaped region*. Underline *crescent-shaped*, point to the hyphen, and explain that when two or more words describe the same noun, they are often connected by a hyphen. Elicit from students that *region* is a noun, and that the words *crescent* and *shaped* both help to modify, or describe, that noun. Have students find other compound modifiers in the unit. These phrases may include: *straight-line distance, problem-solving strategy, half-circle hole, cone-shaped building, 8-inch slide, two-tier waterfall,* and *year-round waterfalls*. Discuss the meaning of the phrases as a class.

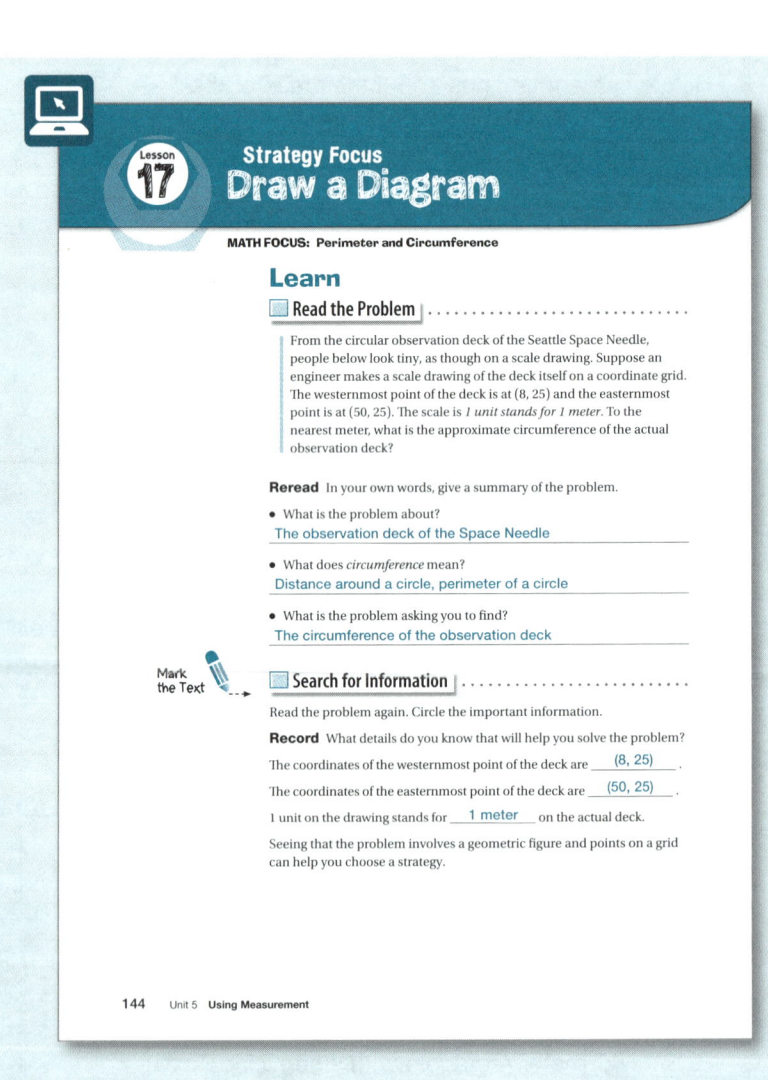

Lesson 17

Strategy Focus: Draw a Diagram

Lesson Overview

Lesson Materials: calculator

Skills to Know	Outcome	Math Vocabulary	eResources www.optionspublishing.com
• Plot points on a coordinate grid • Find the circumference of a circle • Find the perimeter of a two-dimensional shape	Students will recognize that drawing a diagram is an efficient way to solve problems involving perimeter and circumference.	circumference, diameter, perimeter	• Interactive Whiteboard Transparency 17 • Homework, Unit 5 Lesson 17 • Know-Find-Use Table • Grid Paper — 1 cm • Problem-Solving Checklist, also available in the student worktext, page 7

Modeled Instruction

Learn

Ask questions to confirm students' comprehension of the problem's context.

- *What shape is the deck?*
- *How does the engineer's drawing compare to the actual deck?*

As students read the problem again, pose questions to help them identify important information.

- *What information do the phrases* westernmost point *and* easternmost point *give you?*
- *What does the scale* 1 unit stands for 1 meter *mean?*

Use a Graphic Organizer You may wish to use this Think Aloud to demonstrate how to help students sort out important information.

I know the problem is about a scale drawing. I will read the problem again, filling in a Know-Find-Use Table to help me organize the details and see how I might use them to solve the problem. In the Know column, I can write the locations of the points given in the problem westernmost = (8, 25) *and* easternmost = (50, 25). *I can also write the scale of the drawing,* 1 unit : 1 meter, *and the fact that the scale drawing is a circle in the Know column. In the Find column, I can write* The circumference of the actual observation deck. *In the Use column, I can write* C = πd, *the formula for the circumference of a circle.*

I can use this information to solve the problem.

94 Unit 5

Decide What to Do

You know the problem gives you information about two points on a circle. You also know the scale used in the drawing.

Ask How can I find the circumference of the observation deck?

- I can *Draw a Diagram* to see where the points are on the circle.
- I can use the diagram and the scale to find the **diameter** of the actual deck. Then I can calculate the **circumference**.

Use Your Ideas

Step 1 Draw a sketch to show the location of the points and the circle on a coordinate grid. Label the points with their coordinates.

Step 2 Use the diagram to find the diameter of the circle.

The diameter of the circle is __42__ units long, which stands for __42__ meters.

Step 3 Calculate the circumference of the circle. Use 3.14 for π.

$C = \pi d$
$\approx$ __3.14 × 42__
$\approx$ __132__

The circumference of the observation deck is about __132 meters__.

Drawing a Diagram helps you see the size and shape of an object. Then you can visualize the problem and find the information you need.

Review Your Work

Make sure your drawing matches the information in the problem.

Clarify If the two points are accurately drawn, do you need to draw an exact circle in your diagram to solve the problem?

Sample Answer: No, because the sketch helps me see that I can use the two points as endpoints of the diameter. So I only need to find 50 − 8 and use the formula for circumference.

145

Modeled Instruction (continued)

Help students understand how they can use what they know to solve the problem.

- *What do you need to know to find the circumference of a circle?*

Pose questions that give meaning to each step in the solution process.

- *How can you use the diagram to find the diameter of the scale drawing?*
- *Could you have found the circumference of the drawing in units and then converted to meters? Explain.*

Emphasize the importance of checking that the drawing shows the information given in the problem.

- *How do you know that the given points are on a diameter of the circle?*

HOTS Clarify Students' explanations should show an understanding that they only need the two points to be accurately drawn to solve the problem.

Try

Solve the problem.

1. The Liberty Bell hangs from a wooden yoke. The yoke is about 45 inches wide and 20 inches high. Suppose a steel cable is placed around the perimeter of the yoke. Assume the cutout is a half circle with a diameter of 27 inches. To the nearest inch, how long would the cable need to be?

Read the Problem and Search for Information

Identify what information is given and what you are asked to find.

Decide What to Do and Use Your Ideas

You know the width of the yoke is __45__ inches and its height is __20__ inches. The diameter of the cutout is __27__ inches.

Step 1 Draw a diagram and label the dimensions you know.

Step 2 Find the dimensions you do not know. Use 3.14 for π.

Distance around half circle ≈ $\left(\frac{1}{2}\right)(3.14)($ __27__ $)$
= __42.39__ , or about __42__ inches

Step 3 Calculate the perimeter. Use the curved distance.

Perimeter ≈ __20 + 20 + 45 + 9 + 9 + 42__

The cable would need to be about __145__ inches long.

Review Your Work

Make sure you included each distance that makes up the perimeter.

Classify How did you know to use the strategy *Draw a Diagram*?

Sample Answer: The problem asked me to find the distance around an object. A diagram makes it easier to keep track of the dimensions of all the sides.

Ask Yourself: What do I need to know in order to find the distance all the way around a figure?

146 Unit 5 Using Measurement

Guided Practice

Try

1. Prompt students to consider the relationship between the given information and the diagram.

Discuss the context of the problem as well as the key information for solving it.

- *How will labeling a diagram help you solve this problem?*

Have students explain how they completed the diagram and used it to solve the problem.

- *How did you find the distance from the bottom corner to the edge of the cutout?*

Suggest students mark each distance on the diagram as they write it in a sum to find the perimeter to make sure they have not missed any numbers.

HOTS Classify Responses should show that students see that drawing a diagram is helpful in organizing the information in this problem.

Lesson 17 95

Scaffolded Practice
Apply

② Make sure students understand how to sketch a diagram of Pueblo Bonito.
- *What direction will the D be facing?*
- *What two distances make up the perimeter of Pueblo Bonito?*

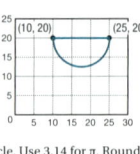

HOTS **Demonstrate** Responses should show that the actual curved part of the perimeter is longer than the curved part of the semicircle.

③ Have students describe how they will label the diagram.
- *How can you find the distance along the front from each corner to the semicircle?*

HOTS **Build** Responses should indicate that labeling all the distances on the diagram helps students keep track of dimensions as they determine them. With a labeled diagram, students can be sure to include all dimensions when finding the perimeter.

④ Make sure students understand that the base of the Statue of Liberty is square. Help them understand how the given coordinates can be used to draw the square.
- *What are the coordinates of the other two corners of the square?*

HOTS **Explain** Students' explanations should mention that drawing a diagram helps them visualize the square base so they can find its perimeter.

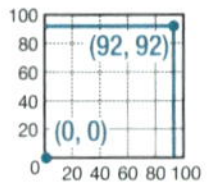

⑤ Help students understand how to draw the diagram.
- *What shape will you draw first? Why?*
- *How will you show the path to the front side of the Lincoln Memorial?*

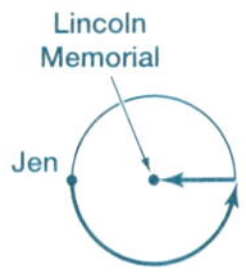

HOTS **Compare** Responses should demonstrate students' ability to analyze problems for similarities and differences.

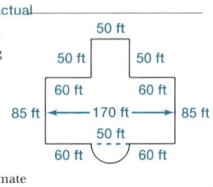

96   Unit 5

Practice

Solve the problems. Show your work.

6 Some buildings live up to their name, like the Pentagon in Washington, D.C., which is shaped like a regular pentagon. On a scale drawing of this building, two adjacent corners are at (13, 0) and (105, 0). The scale is *1 unit stands for 10 feet*. What is the approximate perimeter of the Pentagon?

Some students will use the lesson strategy; however, other strategies may be used. Accept all reasonable work leading to the correct answer.

Answer The perimeter is about 4,600 feet.

Identify This problem includes some hidden information. Explain what you must know about pentagons to solve the problem.

Sample Answer: You need to know that a regular pentagon has 5 sides that are all the same length.

7 Each year, about 13 million people visit Golden Gate Park in San Francisco. Just west of the park's center is a sports area, built over 100 years ago, called Golden Gate Park Stadium. It is shaped like a rectangle, with half circles attached to the shorter sides. The rectangle is 1,030 feet long and 685 feet wide. To the nearest 10 feet, what is the perimeter of the stadium? Use 3.14 for π.

Some students will use the lesson strategy; however, other strategies may be used. Accept all reasonable work leading to the correct answer.

Answer The perimeter of this sports area is about 4,210 feet.

Analyze Why is it not necessary to make a scale drawing of the stadium to answer this question?

Sample Answer: All that is needed to answer the question is a sketch that shows how the dimensions are related.

Create Describe a building or other structure with a combination of shapes, such as rectangles and half circles. Write a problem about your building that can be solved by drawing a diagram. Then draw a diagram and solve your problem.

See teacher notes.

Lesson 17 Strategy Focus: Draw a Diagram 149

Create

In this lesson, students must write a problem that involves a plan for a building. If students are struggling, suggest they first draw a sketch using simple shapes. Then they can label the dimensions.

Accept student responses that include the drawing of a building or other structure on a coordinate grid. The problem should give enough information for someone to duplicate the drawing. The problem should also include the scale used to make the drawing and a correct solution.

Independent Practice
Practice

Students should be encouraged to choose any strategy to solve Problems 6 and 7, though many may prefer to use *Draw a Diagram*.

6 Some students may subtract and use a formula to find the perimeter without drawing a diagram.

Sample Work

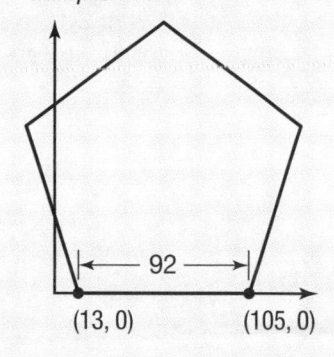

(13, 0)   92   (105, 0)

$P = 5(10s) = 5(10 \times 92) = 4,600$

Identify Responses should indicate that students are familiar with the properties of a regular pentagon.

7 Make sure students understand that the stadium consists of a rectangle with a half circle at each end.

Sample Work

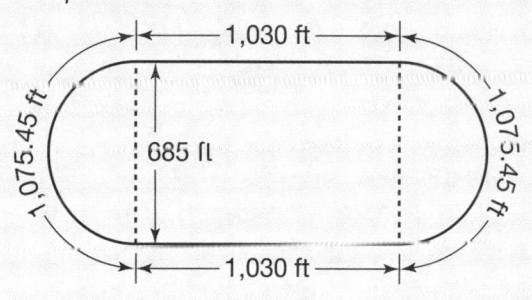

1,030 ft
1,075.45 ft
685 ft
1,075.45 ft
1,030 ft

$\frac{1}{2} \times 3.14 \times 685 = 1,075.45$

$P = 1,030 + 1,075.45 + 1,030 + 1,075.45 = 4,210.9$

Analyze Students' explanations should note that as the problem does not involve scale, a simple sketch showing the dimensions will suffice.

Lesson 17   97

Lesson 18

Strategy Focus
Solve a Simpler Problem

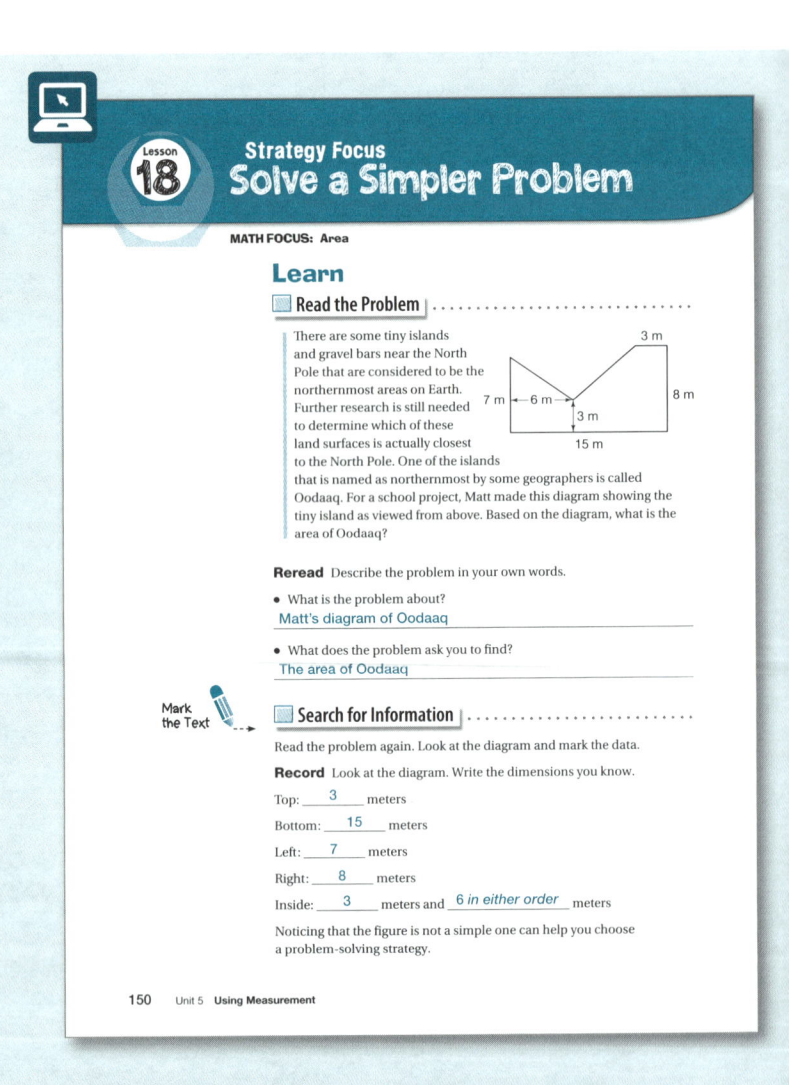

Lesson Overview

Lesson Materials: calculator

Skills to Know	Outcome	eResources www.optionspublishing.com
• Find areas of two-dimensional shapes	Students will recognize that solving a simpler problem is an efficient way to solve problems involving the area of complex shapes.	• Interactive Whiteboard Transparency 18 • Homework, Unit 5 Lesson 18 • Problem-Solving Checklist, also available in the student worktext, page 7

Modeled Instruction

Learn

To be sure students understand the context of the problem, ask questions like the ones below.

- Why do you think Matt made a diagram of Oodaaq rather than describing it in words?
- What units of measure will you use to describe Oodaaq's area?

As students read the problem again, ask questions to help them focus on the details needed to solve it.

- Where can you find the information needed to solve this problem?
- Why do you think only some of the dimensions of Oodaaq are given?

Reread You may wish to use this Think Aloud to demonstrate how to read a diagram for different purposes.

I know the problem is about finding the area of some land. I see a diagram showing the shape of the land. I am going to look at the diagram to see what information is given and think about how I could use that information to solve the problem.

I can see that this shape is not one for which I know an area formula. But I see that this shape can be divided into triangles and rectangles. I do know how to use formulas to find the areas of these shapes. Maybe I can find the areas of all the small shapes, then add them together to find the area of the large shape.

So let me first divide the shape into rectangles and triangles. Wait, I think I am missing some of the dimensions I need to find the areas of all the smaller shapes. I'll see if I can use the dimensions that are given to find the missing dimensions.

Lesson 18 — Strategy Focus: Solve a Simpler Problem

MATH FOCUS: Area

Learn

Read the Problem

There are some tiny islands and gravel bars near the North Pole that are considered to be the northernmost areas on Earth. Further research is still needed to determine which of these land surfaces is actually closest to the North Pole. One of the islands that is named as northernmost by some geographers is called Oodaaq. For a school project, Matt made this diagram showing the tiny island as viewed from above. Based on the diagram, what is the area of Oodaaq?

Diagram dimensions: 3 m (top), 7 m (left inside), 6 m, 3 m, 8 m (right), 15 m (bottom)

Reread Describe the problem in your own words.

- What is the problem about?
  Matt's diagram of Oodaaq
- What does the problem ask you to find?
  The area of Oodaaq

Search for Information

Read the problem again. Look at the diagram and mark the data.

Record Look at the diagram. Write the dimensions you know.

Top: ___3___ meters
Bottom: ___15___ meters
Left: ___7___ meters
Right: ___8___ meters
Inside: ___3___ meters and ___6___ meters *(in either order)*

Noticing that the figure is not a simple one can help you choose a problem-solving strategy.

150   Unit 5   Using Measurement

---

98   Unit 5

Decide What to Do

The diagram shows that Oodaaq is in the shape of a complex figure. You also know some of its dimensions.

Ask How can I find the area of Oodaaq?

- I can use the strategy *Solve a Simpler Problem* to divide the complex figure into simpler shapes.
- I can use the formulas I know to find the areas of the simple shapes. Then I can add to find the total area.

Use Your Ideas

Step 1 Visualize the figure as a combination of triangles and rectangles.

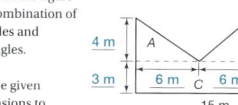

Step 2 Use the given dimensions to determine the dimensions of the simpler shapes and label them.

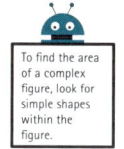

To find the area of a complex figure, look for simple shapes within the figure.

Step 3 Use formulas to calculate the areas of the simpler shapes. Then add to find the area of the complex figure.

$A = \frac{1}{2} \times \underline{4} \times 6 = \underline{12}$ square meters

$B = \frac{1}{2} \times \underline{5} \times \underline{6} = \underline{15}$ square meters

$C = \underline{3} \times 15 = \underline{45}$ square meters

$D = 3 \times \underline{5} = \underline{15}$ square meters

Based on the diagram, the area of Oodaaq is $\underline{87}$ square meters.

Review Your Work

To check if your answer is reasonable, estimate by finding what the area would be if the figure were a 7 × 15 rectangle.

Describe How did the strategy *Solve a Simpler Problem* help you solve this problem?
Sample Answer: There is no single formula to use to find the area of a complex figure. So I had to divide the complex figure into simpler shapes. I found the areas of the simpler shapes and added to find the area of the complex figure.

151

Try

Solve the problem.

1. Wonderwerk Cave shows evidence of human occupation from 2 million years ago. Ashley is designing a diorama of the cave. To the nearest square meter, what is the approximate area of the shaded region shown? Use 3.14 for π.

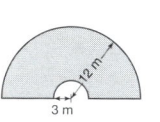

Read the Problem and Search for Information

Reread, then circle the key data in the problem and the diagram.

Decide What to Do and Use Your Ideas

You can use the strategy *Solve a Simpler Problem* to find the area.

Step 1 Visualize the region as a combination of simpler shapes.
The diagram looks like a half circle with a half-circle hole.

Step 2 Determine the dimensions of the simpler figures and label them.
The radius of the large circle is $\underline{12}$ meters.
The radius of the smaller circle is $\underline{3}$ meters.

Step 3 Find the areas of the simpler figures and subtract. The formula for the area of a circle is $A = \pi r^2$. For a half circle, use $A = \frac{1}{2}\pi r^2$.

Large area $= \frac{1}{2}\pi(12^2)$ Small area $= \frac{1}{2}\pi(3^2)$
$= \frac{1}{2}(3.14)(144)$ $= \frac{1}{2}(3.14)(9)$
≈ 226.08 ≈ 14.13

The area of the shaded region is about $\underline{212}$ square meters.

Review Your Work

To check if your answer is reasonable, compare it to the area of a rectangle that would just enclose the shaded region.

Compare How was solving this problem like solving the Learn problem? How was it different?
Sample Answer: Both solutions involved visualizing a complex figure as a combination of simpler figures. In the first problem, I added. They're different because in this problem, I subtracted.

Mark the Text

Ask Yourself What simpler figures make up this figure?

152 Unit 5 Using Measurement

Modeled Instruction (continued)

Help students make a connection between what they know and what they need to find out.

- *How do you find the area of two-dimensional shapes such as rectangles or triangles?*
- *How can you use the area formulas you know to find the area of Oodaaq?*

Pose questions that help students focus on the steps used to solve the problem.

- *Is there more than one way to divide the figure? Why do you think this way was used?*
- *How can you find the height of each triangle?*

Emphasize the importance of checking that the area of the complex figure is reasonable.

- *Should your answer be lesser or greater than your estimate? Why?*

HOTS Describe Explanations should show that students see how to solve and combine simpler area problems to solve problems involving complex figures.

Guided Practice

Try

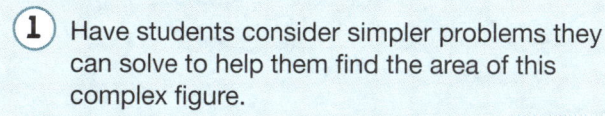

1. Have students consider simpler problems they can solve to help them find the area of this complex figure.

Make sure students understand the information and context of the problem.

- *How would you describe Ashley's diorama?*

Have students explain how they used simpler problems to solve the problem.

- *How do you find the area of each half circle? Why do you subtract?*

Have students explain how their estimate should compare to the area of the shaded region.

HOTS Compare Responses should show that students understand the similarities and differences between how to solve the two problems, including knowing when the areas of simpler figures need to be added and when they need to be subtracted to solve the problem.

Lesson 18 99

Scaffolded Practice
Apply

② Students may be distracted by the fact that the two circles have different centers. To help students solve this problem, ask questions like these.

- What is the radius of the smaller circle? The larger circle?
- Will you add or subtract to find the area of the shaded part? Why?

HOTS Conclude Responses should show an understanding of the relationship between the exact answer and the estimate.

③ Guide students to use the diagram to find the dimensions of the shapes.

- What are the dimensions of the outer parallelogram? The inner parallelogram?

HOTS Explain Students' explanations should indicate an understanding of how to divide a complex figure into simpler figures and how to make sure each part is counted exactly once.

④ Make sure students understand they can use the given dimension to help them find other dimensions needed to solve the problem.

- How can you find the length of each side of the square?

HOTS Choose Explanations should include a description of how to make a useful estimate.

⑤ Ask students to identify the steps they will take to solve the problem.

- Once you find the areas of the simpler figures that make up the harbor and rock formations, what steps will you take? Why?

HOTS Apply Responses should show that students know how to find a percent and which values represent the part versus the whole. In this case, the part is the area of the rock formation and the whole is the area of the entire harbor.

100   Unit 5

Apply

Solve the problems.

② The Great Barrier Reef is the world's largest coral reef. Lisa drew the circles at the right to estimate the area of a small portion of the reef. The radius of the larger circle is 25 meters. To the nearest square meter, what is the approximate area of the shaded part of the figure? Use 3.14 for π.

◀ **Hint** Think of the crescent-shaped region as what is left when one circle is cut out of another.

Area of large circle = $\pi r^2 \approx (3.14)(\underline{\quad 25 \quad})^2$
$\approx 1,962.5$

Area of small circle = $\pi r^2 \approx (3.14)(\underline{\quad 20 \quad})^2$
$\approx 1,256$

Ask Yourself How can I turn this problem into a problem with simpler figures?

Answer The area of the shaded part of the figure is about 707 square meters.

Conclude Would finding the area of half of the large circle be a useful estimate? Why or why not?

Sample Answer: Yes, because the shaded area looks less than half the area of the large circle. So my answer should be less than half of 1,962.5 square meters.

③ Kanton Island is an atoll in the tropical Pacific. It is a strip of land approximately 5 meters wide that surrounds a lagoon. What is the area of the atoll, which is represented by the unshaded region in this diagram?

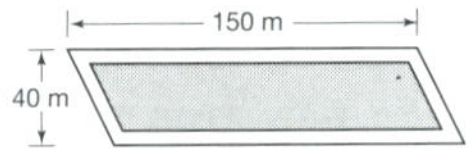

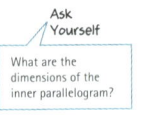

Ask Yourself What are the dimensions of the inner parallelogram?

Area of outer parallelogram = $\underline{\quad 6,000 \quad}$ square meters
Area of inner parallelogram = $\underline{\quad 4,200 \quad}$ square meters

◀ **Hint** The formula for the area of a parallelogram is base × height.

Answer The area of the atoll is 1,800 square meters.

Explain How could you solve this problem by using the *Solve a Simpler Problem* strategy with addition instead of subtraction?

Sample Answer: I could divide the unshaded region into 4 parallelograms and then add their areas to the total area.

Lesson 18   **Strategy Focus: Solve a Simpler Problem**   153

---

④ Some irrigation systems are used to water circles where plants grow. Plants do not grow in the regions between the circles. In the square piece of land represented in the diagram, what is the approximate total area of the land where plants do not grow? Use 3.14 for π.

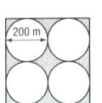

Ask Yourself How many diameters make up one side of the square?

Area of square = $\underline{\quad 160,000 \quad}$ square meters
Area of one circle ≈ $\underline{\quad 31,400 \quad}$ square meters

Hint Be sure to use all 4 circles. ▶

Answer The area of land where plants do not grow is about 34,400 square meters.

Choose How could you estimate to see if your answer is reasonable?
Sample Answer: About $\frac{1}{4}$ of the square looks like it is shaded. A good estimate would be $\frac{1}{4}$ of the area of the square, which is 40,000 square meters.

⑤ This figure is a view, from directly above, of a harbor with a pointed rock formation. The rock formation sticks out of the water. Based on the diagram, what is the approximate area of the surface of the water? Use 3.14 for π.

Ask Yourself What common shapes do I see?

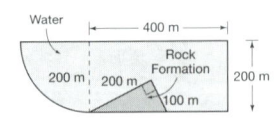

Hint The area of a quarter circle is $\frac{1}{4}$ the area of the whole circle. ▶

Total area shown ≈ $\underline{\quad 111,400 \quad}$ square meters
Area of rock formation = $\underline{\quad 10,000 \quad}$ square meters

Answer The area of the surface of the water is about 101,400 square meters.

Apply If you needed to find the percentage of the harbor covered by the rock formation, what simple problem could you solve?

Sample Answer: I could divide the area of the rock formation by the harbor area, including the rock formation.

154   Unit 5   Using Measurement

Practice

Solve the problems. Show your work.

6) Champagne Pool in New Zealand got its name because it releases bubbles of carbon dioxide. An orange strip caused by natural chemicals encircles the pool. The pool is in a circular crater that is about 62 meters in diameter. An orange ring about 1 meter wide lies on the inside edge of the crater. To the nearest square meter, what is the area of the orange ring? Use 3.14 for π.

Some students will use the lesson strategy; however other strategies may be used. Accept all reasonable work leading to the correct answer.

Answer The area of the orange ring is about 192 square meters.

Determine Suppose you multiply the circumference of the inner circle by the width of the orange ring. How close is the result to your answer? Explain.

Sample Answer: It's very close. It's only π square meters less.

7) In Africa, there are extinct volcanic craters with lush grazing land and watering holes. The figure here is inspired by one of these craters. The watering hole has the shape of a trapezoid. The grazing land has the shape of the circle shown, but does not include the watering hole. What is the area of the grazing land? Use 3.14 for π.

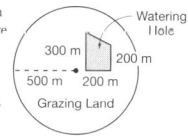

(Figure: Watering Hole trapezoid with 300 m, 200 m, 500 m, 200 m dimensions; Grazing Land circle)

Some students will use the lesson strategy; however other strategies may be used. Accept all reasonable work leading to the correct answer.

Answer The area of the grazing land is about 735,000 square meters.

Justify If you subtract the area of a rectangle that is 200 × 300 meters from the area of the circle, would you get the right answer? Explain.

Sample Answer: No, you need to subtract the area of the trapezoid, which is 10,000 square meters smaller than the rectangle.

Create Write and solve a problem about finding the area of a complex figure that can be divided into triangles and rectangles. Be sure you can solve the problem using the strategy *Solve a Simpler Problem*.

See teacher notes.

Lesson 18 Strategy Focus: Solve a Simpler Problem   155

Independent Practice
Practice

Students should be encouraged to choose any strategy to solve Problems 6 and 7, though many may prefer to use *Solve a Simpler Problem*.

6) Some students may choose to draw a diagram to help solve the problem. Be sure they understand the ring is inside, not outside, the crater.

Sample Work

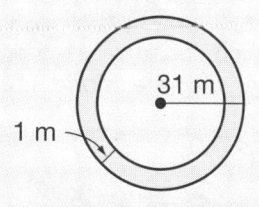

(Diagram: concentric circles with 31 m radius, 1 m ring width)

Area of crater ≈ $(3.14)(31)^2$
 = 3,017.54
Area of inner circle ≈ $(3.14)30^2$
 = 2,826
$3,017.54 - 2,826 = 191.54$

HOTS Determine Responses should show how students made their determinations that the results are close.

7) If students remember the formula, they may choose to find the area of the trapezoid directly.

Sample Work

Area of square = $(200)(200) = 40,000$
Area of triangle = $\frac{1}{2}(200)(100) = 10,000$
Area of trapezoid = $40,000 + 10,000 = 50,000$
Area of circle ≈ $(3.14)(500)^2 = 785,000$
$785,000 - 50,000 = 735,000$

HOTS Justify Students' responses should indicate an understanding that they need to find the area of the trapezoid and subtract that from the area of the circle.

Create In this lesson, students write a problem about the area of a complex figure. If students are struggling, suggest they look back at the first problem in the lesson for ideas.

Accept student responses that include a complex figure that can be divided into triangles and rectangles. Enough dimensions should be given to enable the student to find the areas of all the triangles and rectangles that make up the complex figure. Responses should include a correct solution.

Lesson 18   101

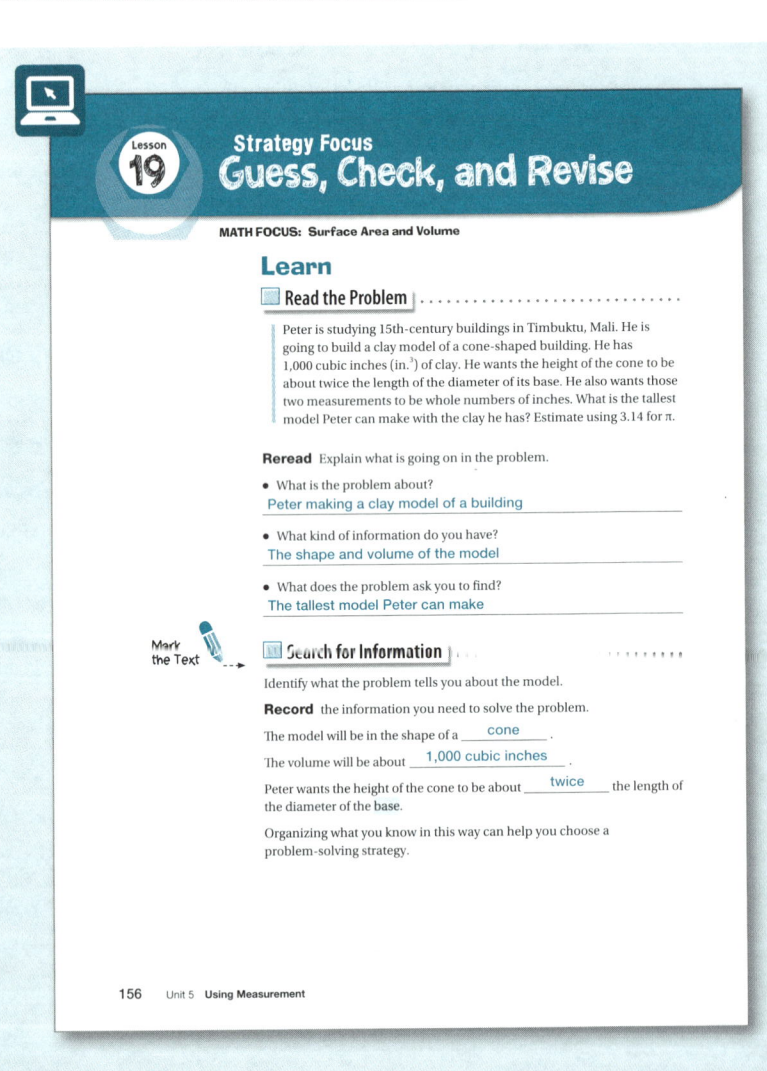

Lesson 19

Strategy Focus: Guess, Check, and Revise

Lesson Overview

Lesson Materials: calculators

Skills to Know	Outcome	Math Vocabulary	eResources www.optionspublishing.com
• Find volumes of three-dimensional shapes • Apply the Pythagorean Theorem	Students will recognize that guessing, checking, and revising is an efficient way to solve problems in which conditions, rather than discrete information, are given.	base, prism, pyramid	• Interactive Whiteboard Transparency 19 • Homework, Unit 5 Lesson 19 • Know-Find Table • Problem-Solving Checklist, also available in the student worktext, page 7

Modeled Instruction

Learn

Probe students' understanding of the problem's context by asking questions similar to the following.

- *Why might Peter want the height and diameter of his model to be whole numbers of inches?*

As students read the problem again, guide them to identify the words and numbers needed to solve the problem.

- *Do you think Peter will use all of the clay to build his model? Explain.*

Use a Graphic Organizer You may wish to use this Think Aloud to demonstrate how to organize information and identify special conditions.

I know the problem is about the volume of a model. I will read the problem again, filling in the Know-Find Table to help me organize the details to see how I might use them.

In the Know column, I will write The model will be cone-shaped *and* The volume of the clay is 1,000 cubic inches. *I can also write* The height is twice the diameter of the base *and* The measurements should be whole inches.

In the Find column, I will write The tallest Peter can make his model.

I wonder if there is a way I can use the formula for the volume of a cone to help me find a solution to the problem?

102   Unit 5

Decide What to Do

You know the shape and volume of the model, the relationship between the height and the diameter, and the type of numbers that can be used.

Ask How can I find the tallest model Peter can make?

- I can use the strategy *Guess, Check, and Revise* to guess different radii and heights and calculate the volume.
- I can revise incorrect guesses to make the next guess closer.

Use Your Ideas

Step 1 Use the table. Guess a radius and height using whole numbers. Check to see if those guesses result in the correct volume.

Since the height will be twice the diameter, the height will be __4__ times the radius.

Step 2 Revise your guess so you get closer to the correct result. Continue until your answer matches the conditions.

radius, r	height, h	$V = \frac{1}{3}\pi r^2 h$	Is V ≈ 1000?
5 in.	20 in.	$\frac{1}{3}(3.14)(5^2)(20)$	523.33; no
10 in.	40 in.	$\frac{1}{3}(3.14)(10^2)(40)$	4,186.67; no
7 in.	28 in.	$\frac{1}{3}(3.14)(7^2)(28)$	1,436.03; no
6 in.	24 in.	$\frac{1}{3}(3.14)(6^2)(24)$	904.32; yes

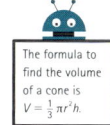

Students' guesses will vary.

The formula to find the volume of a cone is $V = \frac{1}{3}\pi r^2 h$.

The tallest model Peter can make with the clay he has is __24__ inches tall.

Review Your Work

Check that your answer meets the conditions given in the problem.

Consider How did each guess help you find the answer?

Sample Answer: I used the information from each guess to get closer to the correct solution.

157

Modeled Instruction (continued)

Help students make a connection between the facts they know and what strategy they can use.

- *How are the different measurements related?*
- *How can you check that the numbers you guessed are correct?*

Ask questions that guide students to consider each step in the solution process.

- *Why is the height four times the radius?*
- *Which of the first two guesses is closer to the correct answer? How might knowing this help?*

Emphasize the importance of checking that the answer meets all of the given conditions.

- *How can you be sure that your answer gives a volume that is closest to 1,000 in.³ without going over?*

HOTS Consider Students' explanations should recognize that an incorrect guess helps them make better subsequent guesses.

Try

Solve the problem.

① Jorge is making a model inspired by the Mayan pyramid El Castillo in Chichén Itzá, Mexico. His plan shows a side view of the model. According to his plan, what will the height of the model be?

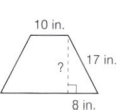

10 in.
17 in.
?
8 in.

Read the Problem and Search for Information

Reread, then circle the important data in the problem.

Decide What to Do and Use Your Ideas

You can use the strategy *Guess, Check, and Revise* and the Pythagorean Theorem to find the missing height. The figure made by the height, the 8-inch side, and the 17-inch side is a right triangle.

Step 1 Use the table. Guess the height. Use the Pythagorean Theorem to check whether that height gives the hypotenuse of 17 inches.

$a = $ __8__ , $b = $ height, $c = $ __17__ , and $c^2 = $ __289__

Step 2 Revise and guess again, based on what you learned from your previous guess. Continue until your answer matches the conditions.

a	a²	b	b²	Does a² + b² = 289?
8	64	10	100	164; no
8	64	20	400	464; no
8	64	15	225	289; yes

Students' guesses may vary.

The height of the model will be __15 inches__.

Review Your Work

Use the diagram. Estimate to check that your answer is reasonable.

Explain If your first guess is 10 inches for the height, why would trying 5 inches next not be a useful guess?

Sample Answer: I should guess a number greater than 10.

$8^2 + 10^2$ is less than 289, so $8^2 + 5^2$ will be even less.

Unit 5 Using Measurement
158

Guided Practice

Try

① Prompt students to consider how the information given relates to what information they need to find.

Discuss the context of the problem as well as the key information needed to solve it.

- *What part of the plan represents the height of the pyramid?*

Guide students to recognize which elements in the table change and which remain the same.

- *Why don't the values for a and a² change?*
- *How does the table help you find a solution?*

Have students explain how they could substitute the given dimensions for a and c in the Pythagorean Theorem to check their measurement for b.

HOTS Explain Explanations should note that the first guess was too low, so subsequent guesses should be greater than 10.

Lesson 19   103

Scaffolded Practice
Apply

2 Help students recognize the relationships among the measurements in the problem.
- After you guess the depth, how do you find the height?
- Should your next guess be greater than or less than 5 inches? How did you decide?

Evaluate Explanations should note that *Guess, Check, and Revise* is an effective strategy to use when all of the necessary facts are not given, but the conditions relating the facts are given.

3 Help students relate the measurement they are guessing to what they are asked to find.
- After you find a radius that gives a volume close to 1,256 in.³, what do you do next? Why?

Discuss Responses should indicate the need to keep work organized to guide subsequent guesses.

4 Make sure students understand that the dashed line showing the height of the isosceles triangle divides the triangle into two congruent right triangles.
- How will you use values in the table to answer the question?

Analyze Explanations should recognize that *a* is half the length of the missing side length of the base.

5 Help students make reasonable guesses.
- What factors will you consider as you make the first guess?
- How will your results guide your next guess?

Plan Responses should indicate that students knew they had found the solution when they found the two whole numbers that the radius was between. At that point, they just needed to choose the guess that gave a closer result to the actual volume.

Apply

Solve the problems.

2 Pia has 270 cubic inches of clay to make a model of a Mayan ruin. She will use all of the clay. The height must be 3 times the depth. How high and how deep will the model be?

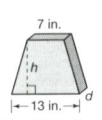

depth, d	height, h	V = $\frac{1}{2}$(13 + 7)hd	Does V = 270?
5 in.	15 in.	$\frac{1}{2}$(20)(5)(15)	750; no
3 in.	9 in.	$\frac{1}{2}$(20)(9)(3)	270; yes

Students' guesses may vary.

Answer The block will be 3 inches deep and 9 inches high.

Evaluate Why is *Guess, Check, and Revise* an effective strategy for this problem?

Sample Answer: The problem gives conditions that must be met, so you can tell if an answer is correct if it meets those conditions.

Hint The formula for the volume of a trapezoidal prism is V = $\frac{1}{2}$ × (sum of the base lengths) × height × depth.

Ask Yourself If I guess that d is 2, what must h be for that guess?

3 Paula will use a cylinder of clay with a volume of 1,256 cubic inches and a height of 16 inches to model an Easter Island statue. To the nearest inch, what is the diameter of the cylinder? Use 3.14 for π.

radius, r	height, h	V = πr²h	Is V ≈ 1,256?
10 in.	16 in.	(3.14)(10²)(16)	5,024; no
5 in.	16 in.	(3.14)(5²)(16)	1,256; yes

Students' guesses may vary.

Answer The diameter of the cylinder is 10 inches.

Hint The formula for the volume of a cylinder is: V = πr²h.

Ask Yourself How is the diameter of a cylinder related to its radius?

Discuss How is it helpful to you to use a table when making guesses?

Sample Answer: A table helps me to organize my guesses and all the steps I need to follow to check them.

Lesson 19 Strategy Focus: Guess, Check, and Revise 159

Ask Yourself How can I use a² + b² = c² to help me relate the missing information to the given information?

Hint The dashed 8-centimeter line on Marne's diagram forms two right triangles.

4 After studying the Great Pyramid of Giza, Marne decides to make a model of it by making a net of a square pyramid. It looks like he forgot to write down some of the measurements on his diagram. What is the length of each side of the base?

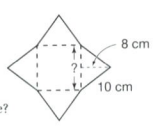

a	a²	b	b²	Does a² + b² = 100?
3 cm	9	8 cm	64	73 no
6 cm	36	8 cm	64	100 yes

Students' guesses may vary.

Answer The length of a side of the square base is 12 cm.

Analyze What does *a* represent in the pyramid?

Sample Answer: a is half the length of a side of the square base of the pyramid.

Ask Yourself How can I make a first guess for the radius?

5 For a school project, Pam researched the size of the Tower of Pisa and made notes of some approximate dimensions. Her notes are smudged. To the nearest meter, what is the approximate radius of the Tower of Pisa? Use 3.14 for π.

Tower of Pisa
Shape: Cylinder
Volume: 10,540 m³
Height: 56 m
Diameter: ▮

Hint Small changes in the radius length have a big effect on the volume because when you calculate the volume, you square the length of the radius.

radius, r	V = πr²h	Is V ≈ 10,540?
5 m	(3.14)(5²)(56)	4,396 no
8 m	(3.14)(8²)(56)	11,254 yes
7 m	(3.14)(7²)(56)	8,616 no

Student guesses may vary.

Answer The radius is about 8 meters long.

Plan How did you know when you found the solution?

Sample Answer: I knew when two consecutive numbers for radii gave an answer close to the volume. So I picked the closer one.

160 Unit 5 Using Measurement

Practice

Solve the problems. Show your work.

6 Inspired by intricate tiling in the Taj Mahal, Wan is designing a tile pattern using rectangular tiles. He wants the diagonal of each rectangle to be 13 centimeters and the width 5 centimeters. What is the length of the rectangle?

Some students will use the lesson strategy; however, other strategies may be used. Accept all reasonable work leading to the correct answer.

Answer This rectangle is 12 centimeters long.

Compare How is this problem like Problem 1?

Sample Answer: Solving both involves using the Pythagorean Theorem to find a missing length.

7 Emily is building a model of part of the Great Wall of China. The height of that part of the wall is about $1\frac{1}{2}$ times its depth. Her wall is 60 centimeters long and is made of 3,240 cubic centimeters of clay. How tall is Emily's model?

Some students will use the lesson strategy; however, other strategies may be used. Accept all reasonable work leading to the correct answer.

Answer Her model is 9 centimeters tall.

Evaluate Why is it reasonable to solve this problem by guessing and checking?

Sample Answer: I know a relationship between the height and the width, but I don't know the actual measures.

Create Create a problem like Problem 5, choosing your own volume and height. Write and solve the problem. Be sure your problem can be solved using the strategy *Guess, Check, and Revise*.
See teacher notes.

Lesson 19 Strategy Focus: Guess, Check, and Revise   161

Create

In this lesson, students solved measurement problems in which at least one dimension was not given. Now they will write a problem like Problem 5. If students are struggling, suggest they first draw a diagram of a cylinder that shows the height and diameter. Then have them find the volume and write a problem using that answer, omitting the diameter.

Accept responses that give the volume and height of a cylinder. The problem should ask for the radius or diameter of the cylinder and should include a correct solution using the strategy *Guess, Check, and Revise*.

Independent Practice
Practice

Students should be encouraged to choose any strategy to solve Problems 6 and 7, though many may prefer to use *Guess, Check, and Revise*.

6 Some students may choose to write and solve an equation using the Pythagorean Theorem to solve the problem directly, without using *Guess, Check, and Revise*.

Sample Work

$a^2 + b^2 = c^2$

$c^2 = 13^2 = 169$

a	a^2	b	b^2	Does $a^2 + b^2 = 169$?
10	100	5	25	125 no
15	225	5	25	250 no
12	144	5	25	169 yes

HOTS Compare Responses should indicate that students recognize that both problems give the dimensions of two sides of a right triangle and the third side can be found using the Pythagorean Theorem.

7 Make sure students understand that Emily's model is a rectangular prism.

Sample Work

l	w	h (w × 1.5)	Does *lwh* = 3,240?
60	4	6	1,440 no
60	8	12	5,760 no
60	6	9	3,240 yes

HOTS Evaluate Students' explanations should mention that *Guess, Check, and Revise* is an efficient strategy to use when some measurements are unknown, but their relationships to known measurements are given.

Lesson 19   105

Lesson 20

Strategy Focus: Write an Equation

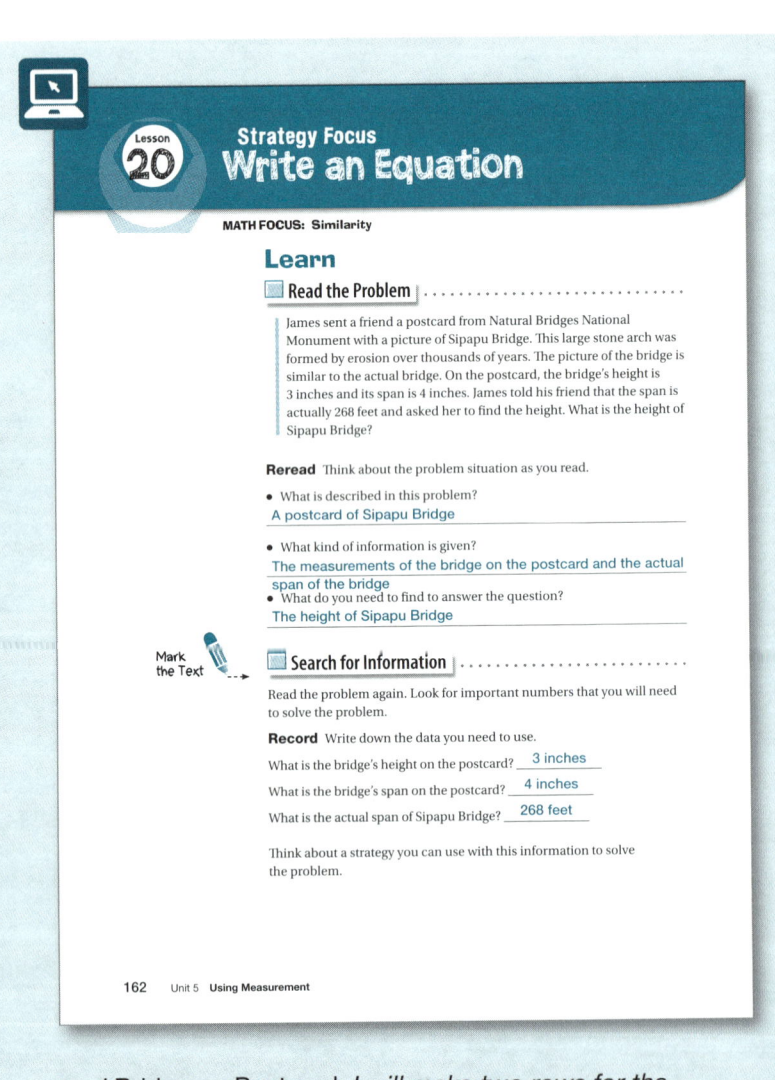

Lesson Overview

Lesson Materials: calculator

Skills to Know	Outcome	Math Vocabulary	eResources www.optionspublishing.com
• Set up and solve proportions	Students will recognize that writing an equation is an efficient way to solve problems involving similar figures.	equivalent ratios, proportion	• Interactive Whiteboard Transparency 20 • Homework, Unit 5 Lesson 20 • Problem-Solving Checklist, also available in the student worktext, page 7

Modeled Instruction

Learn

Ask questions about the problem's context to clarify students' comprehension of what the problem is about.

- *How are the actual bridge and the bridge on the postcard the same? How are they different?*
- *What does the span of a bridge refer to?*

As students read the problem again, pose questions to help them recognize important phrases and facts.

- *Are there any numbers or number words in the problem that you do not need to solve the problem? How did you decide?*
- *Is the actual height of the bridge greater than or less than the span? How do you know?*

Represent You may wish to use this Think Aloud to demonstrate how a table can be used to set up a proportion.

I know this problem is about an actual bridge and a picture of that bridge on a postcard. I am given some of the dimensions of the bridge on the postcard. I need to find the height of the actual bridge.

I know that the actual bridge and the picture of the bridge are the same shape but different sizes. I think that means the bridges are similar figures. I know the dimensions of similar figures are proportional to each other. So I can write equivalent ratios using the dimensions I am given.

I will reread the problem and organize the information given and what I need to find in a table. I will make two columns for the table: Actual Bridge and Bridge on Postcard. I will make two rows for the table: Height and Span. After I fill in the table, I can use the information in the columns to write two equivalent ratios in a proportion.

Lesson 20 — Strategy Focus: Write an Equation

MATH FOCUS: Similarity

Learn

Read the Problem

James sent a friend a postcard from Natural Bridges National Monument with a picture of Sipapu Bridge. This large stone arch was formed by erosion over thousands of years. The picture of the bridge is similar to the actual bridge. On the postcard, the bridge's height is 3 inches and its span is 4 inches. James told his friend that the span is actually 268 feet and asked her to find the height. What is the height of Sipapu Bridge?

Reread Think about the problem situation as you read.

- What is described in this problem?
  A postcard of Sipapu Bridge
- What kind of information is given?
  The measurements of the bridge on the postcard and the actual span of the bridge
- What do you need to find to answer the question?
  The height of Sipapu Bridge

Mark the Text

Search for Information

Read the problem again. Look for important numbers that you will need to solve the problem.

Record Write down the data you need to use.

What is the bridge's height on the postcard? __3 inches__
What is the bridge's span on the postcard? __4 inches__
What is the actual span of Sipapu Bridge? __268 feet__

Think about a strategy you can use with this information to solve the problem.

162 Unit 5 Using Measurement

Decide What to Do

You know the height and span of the bridge on the postcard and the actual span of Sipapu Bridge. You know that the bridge in the picture is similar to the actual bridge.

Ask How can I find the actual height of Sipapu Bridge?

- I know that the ratios of the spans and the heights of the bridge in the picture and of the actual bridge are equivalent.
- I can use equivalent ratios to *Write an Equation*.

Use Your Ideas

Step 1 Describe the proportion in words first.
$$\frac{\text{Actual height}}{\text{Actual span}} = \frac{\text{Height on postcard}}{\text{Span on postcard}}$$

Step 2 Now use the information you have to write the proportion.

Let h be the height of the bridge, in feet.
$$\frac{h}{268 \text{ feet}} = \frac{3 \text{ inches}}{4 \text{ inches}}$$

Since this is a proportion, the ratio of the measurements in inches is the same as the ratio of those measurements in feet.

Step 3 Simplify and solve the proportion.
$$\frac{h}{268} = \frac{3}{4}$$
$$4h = \underline{268} \times 3$$
$$4h = \underline{804}$$
$$\frac{4h}{4} = \frac{804}{4}$$
$$h = \underline{201}$$

The actual height of Sipapu Bridge is __201 feet__.

Review Your Work

Be sure that you used the correct unit of measure for your answer.

Explain Why is writing an equation a useful strategy for solving this problem?

Sample Answer: I know that the picture of the bridge is similar to the real bridge, so I can relate the information with a proportion.

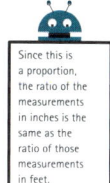

163

Try

Ask Yourself: How can I write two equivalent ratios so I can use a proportion?

Mark the Text

Solve the problem.

① At Yosemite Valley, Inga learns that El Capitan is the world's largest granite monolith. She notices that its shadow is 25 meters long and decides to estimate how high it rises from the valley. She finds that a nearby tree is 20 meters tall and casts a shadow that is 50 centimeters long. About how far above the valley floor does El Capitan rise?

Read the Problem and Search for Information

Identify the information you need to answer this question.

Decide What to Do and Use Your Ideas

You can use the strategy *Write an Equation* to solve a proportion.

Step 1 Convert the units of the tree's shadow to meters, because the other measurements are in meters.
$$50 \text{ cm} = \underline{0.5} \text{ m}$$

Step 2 Use words to write two equivalent ratios as a proportion. Then substitute in the information you have.
$$\frac{\text{El Capitan's height}}{\text{El Capitan's shadow}} = \frac{\text{Tree's height}}{\text{Tree's shadow}}$$

Let h be the height of El Capitan, in meters.
$$\frac{h}{25} = \frac{20}{0.5}$$

Step 3 Simplify and solve the proportion.
$$\frac{h}{25} = \frac{20}{0.5} \rightarrow 0.5h = 25 \times \underline{20} \rightarrow 0.5h = \underline{500}$$
$$\frac{0.5h}{0.5} = \frac{500}{0.5} \rightarrow h = \underline{1,000}$$

El Capitan rises about __1,000 meters__ above the valley floor.

Review Your Work

Check that you have written the proportion correctly.

State What is another proportion you could use to solve this problem?

Sample Answer: I could write equivalent ratios with the heights and the shadow lengths and then use those ratios to form a proportion.

164 Unit 5 *Using Measurement*

Modeled Instruction (continued)

Help students connect the facts they know with how they can use them to write an equation.

- *How does recognizing that the picture of the bridge and the actual bridge are similar figures help you decide on a strategy for solving this problem?*

Ask questions that encourage students to think critically about the steps in the solution process.

- *Why is it helpful to write the proportion in words first?*
- *What steps do you use to solve a proportion?*

Emphasize the importance of checking that answers include the correct unit of measure.

- *How did you know which unit of measure to include in your answer?*

HOTS Explain Explanations should state why proportions can be used to solve problems involving similar figures.

Guided Practice

Try

① Prompt students to consider how the information about El Capitan and the tree are related.

Discuss the context of the problem as well as the information needed to solve it.

- *What is a monolith? What words give you clues?*
- *How do El Capitan, the tree, and their shadows form similar figures?*

Have students explain how the proportion was set up.

- *Why was it necessary to have matching units in this problem, but not in the previous problem?*

Discuss ways that students could check that they have written true proportions.

HOTS State Students' responses should indicate that the ratios in the proportion may compare two measurements for the same figure or the same measurement for two different figures.

Lesson 20   107

Scaffolded Practice
Apply

2 Once students have extracted essential information, ask them to think about the steps they will take to find and check their solution.
- *How will you set up each ratio so you can write a proportion?*
- *How can you check that your answer is reasonable?*

HOTS Conclude Responses should show that students can identify what information is not needed to solve the problem.

3 Students may be distracted by the different units of measure in this problem. Prompt them to think about the measures in the ratio.
- *Do you need to convert any measurements to solve this problem? Why or why not?*

HOTS Relate Students' explanations should show an understanding that the terms in both ratios in a proportion must be written in the same order.

4 Make sure students use the same order for the terms of the ratios in the proportion.
- *How does the proportion given in words help you to write a proportion using numbers and variables?*

HOTS Identify Students' explanations should indicate an understanding that numbers given in a problem that are not exact will result in an answer that is not exact.

5 Encourage students to estimate an answer before finding the exact answer to this problem.
- *What words and numbers help you decide if the height will be greater than or less than $7\frac{1}{2}$ feet? About how many times greater will the height be?*

HOTS Interpret Responses should indicate that students understand properties of similar triangles.

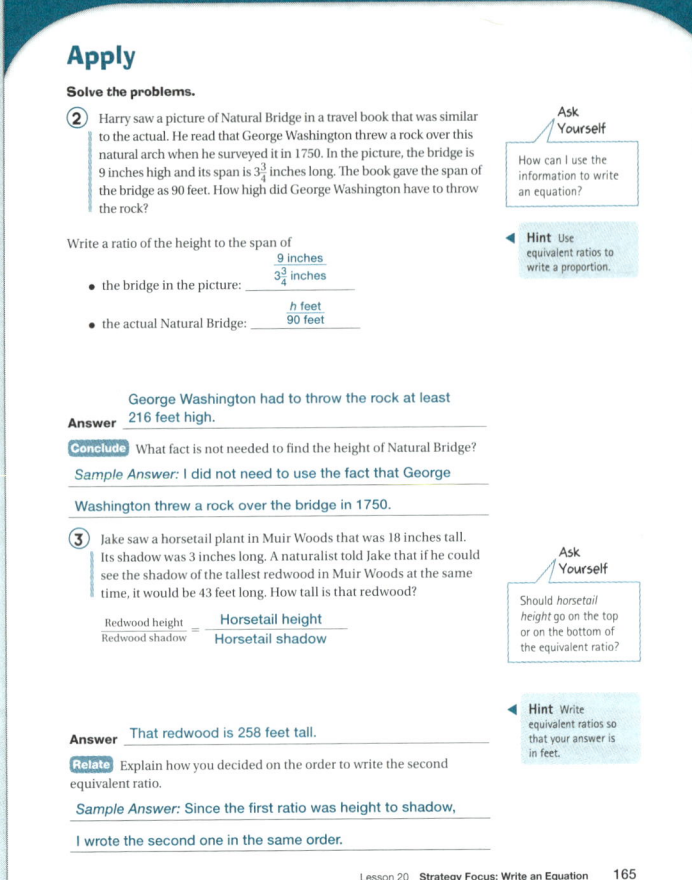

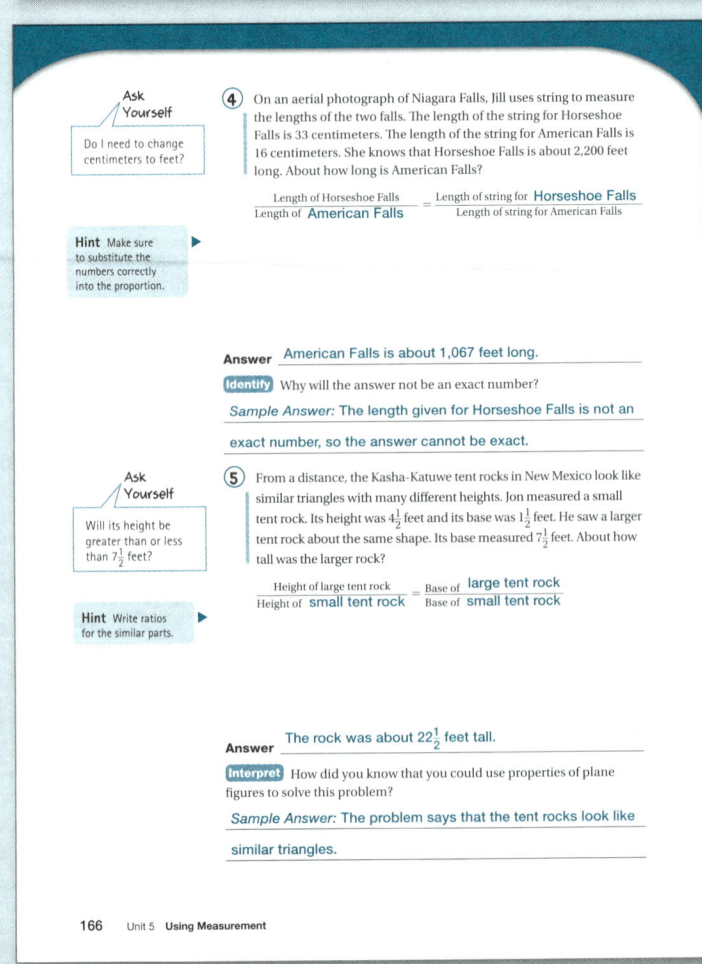

Apply
Solve the problems.

2 Harry saw a picture of Natural Bridge in a travel book that was similar to the actual. He read that George Washington threw a rock over this natural arch when he surveyed it in 1750. In the picture, the bridge is 9 inches high and its span is $3\frac{3}{4}$ inches long. The book gave the span of the bridge as 90 feet. How high did George Washington have to throw the rock?

Ask Yourself: How can I use the information to write an equation?

Hint Use equivalent ratios to write a proportion.

Write a ratio of the height to the span of
- the bridge in the picture: $\dfrac{9 \text{ inches}}{3\frac{3}{4} \text{ inches}}$
- the actual Natural Bridge: $\dfrac{h \text{ feet}}{90 \text{ feet}}$

Answer George Washington had to throw the rock at least 216 feet high.

Conclude What fact is not needed to find the height of Natural Bridge?
Sample Answer: I did not need to use the fact that George Washington threw a rock over the bridge in 1750.

3 Jake saw a horsetail plant in Muir Woods that was 18 inches tall. Its shadow was 3 inches long. A naturalist told Jake that if he could see the shadow of the tallest redwood in Muir Woods at the same time, it would be 43 feet long. How tall is that redwood?

Ask Yourself: Should *horsetail height* go on the top or on the bottom of the equivalent ratio?

$\dfrac{\text{Redwood height}}{\text{Redwood shadow}} = \dfrac{\text{Horsetail height}}{\text{Horsetail shadow}}$

Hint Write equivalent ratios so that your answer is in feet.

Answer That redwood is 258 feet tall.

Relate Explain how you decided on the order to write the second equivalent ratio.
Sample Answer: Since the first ratio was height to shadow, I wrote the second one in the same order.

Lesson 20 Strategy Focus: Write an Equation 165

Ask Yourself: Do I need to change centimeters to feet?

4 On an aerial photograph of Niagara Falls, Jill uses string to measure the lengths of the two falls. The length of the string for Horseshoe Falls is 33 centimeters. The length of the string for American Falls is 16 centimeters. She knows that Horseshoe Falls is about 2,200 feet long. About how long is American Falls?

$\dfrac{\text{Length of Horseshoe Falls}}{\text{Length of American Falls}} = \dfrac{\text{Length of string for Horseshoe Falls}}{\text{Length of string for American Falls}}$

Hint Make sure to substitute the numbers correctly into the proportion.

Answer American Falls is about 1,067 feet long.

Identify Why will the answer not be an exact number?
Sample Answer: The length given for Horseshoe Falls is not an exact number, so the answer cannot be exact.

5 From a distance, the Kasha-Katuwe tent rocks in New Mexico look like similar triangles with many different heights. Jon measured a small tent rock. Its height was $4\frac{1}{2}$ feet and its base was $1\frac{1}{2}$ feet. He saw a larger tent rock about the same shape. Its base measured $7\frac{1}{2}$ feet. About how tall was the larger rock?

Ask Yourself: Will its height be greater than or less than $7\frac{1}{2}$ feet?

$\dfrac{\text{Height of large tent rock}}{\text{Height of small tent rock}} = \dfrac{\text{Base of large tent rock}}{\text{Base of small tent rock}}$

Hint Write ratios for the similar parts.

Answer The rock was about $22\frac{1}{2}$ feet tall.

Interpret How did you know that you could use properties of plane figures to solve this problem?
Sample Answer: The problem says that the tent rocks look like similar triangles.

166 Unit 5 Using Measurement

Practice

Solve the problems. Show your work.

6 Julia saw a picture of Multnomah Falls in a travel book. This two-tier waterfall is one of the tallest year-round waterfalls in the United States. In the picture, the two tiers combined are $8\frac{1}{2}$ inches tall, and the upper falls is $6\frac{1}{4}$ inches tall. Julia read that the combined height of the two tiers is actually 850 feet tall. How tall is the upper falls?

Some students will use the lesson strategy; however, other strategies may be used. Accept all reasonable work leading to the correct answer.

Answer The upper falls is 625 feet tall.

Choose Why can you use part-to-whole equivalent ratios to solve this problem?

Sample Answer: I know the total height of the real falls and the falls in the picture. I know the height of the upper falls in the picture. So I can write ratios of the parts to the wholes.

7 A national park is named for Joshua trees, unusual plants with twisted shapes. Many seem to have similar ratios of height to diameter. Joan saw two Joshua trees in the Mojave Desert. The smaller one was about 8 feet tall and its diameter was about 6 feet across. The larger one had a diameter that was about 15 feet across. About how tall was the larger Joshua tree?

Some students will use the lesson strategy; however, other strategies may be used. Accept all reasonable work leading to the correct answer.

Answer The larger Joshua tree was about 20 feet tall.

Generalize How can drawing quick sketches of the objects in these problems help you set up a proportion?

Sample Answer: The sketches can help me see how all the parts of the problem are related so I can write equivalent ratios.

Create Think about some similar objects you have seen that are shaped like cones, prisms, or other solid figures. Write and solve a problem about two of these objects that can be solved using the strategy *Write an Equation*.

See teacher notes.

Lesson 20 Strategy Focus: Write an Equation 167

Create

In this lesson, students write a problem involving similar figures. Guide students to compare actual dimensions to dimensions in a picture.

Accept student responses that include a problem about two similar objects shaped like three-dimensional figures. The problem should give measurements for two of the dimensions of the first figure. The problem should also give one measurement for the second figure and ask for a missing measurement. Responses should include a correct solution found using *Write an Equation*.

Independent Practice
Practice

Students should be encouraged to choose any strategy to solve Problems 6 and 7, though many may prefer to use *Write an Equation*.

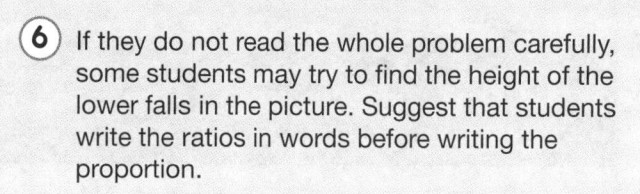

6 If they do not read the whole problem carefully, some students may try to find the height of the lower falls in the picture. Suggest that students write the ratios in words before writing the proportion.

Sample Work

$$\frac{\text{Upper height in picture}}{\text{Combined height in picture}} = \frac{\text{Actual upper height}}{\text{Actual combined height}}$$

$$\frac{6\frac{1}{4}}{8\frac{1}{2}} = \frac{h}{850}$$

$$h = 625$$

HOTS Choose Students explanations should mention that the dimensions are proportional because it is a picture, so equivalent ratios can be used.

7 More than one proportion can be written to solve this problem. Encourage students to check that they use the same order in both ratios.

Sample Work

$$\frac{\text{Height of smaller tree}}{\text{Diameter of smaller tree}} = \frac{\text{Height of larger tree}}{\text{Diameter of larger tree}}$$

$$\frac{8}{6} = \frac{h}{15}$$

$$h = 20$$

HOTS Generalize Students' responses should indicate that sketches can show how the measurements given in a problem are related, which is helpful when writing equivalent ratios.

Lesson 20 109

UNIT 5 Review

UNIT 5 Review

In this unit, you worked with four problem-solving strategies. You can often use more than one strategy to solve a problem. So if a strategy does not seem to be working, try a different one.

Problem-Solving Strategies
- ✓ Draw a Diagram
- ✓ Solve a Simpler Problem
- ✓ Guess, Check, and Revise
- ✓ Write an Equation

Check students' work throughout. Students' choices of strategies may vary.
Solve each problem. Show your work. Record the strategy you use.

1. Betsy is making a quilt out of square pieces of fabric. Each square is 6 inches on a side. The quilt has 7 rows of 5 squares each. Betsy wants to sew a border along the edge of the quilt. How many feet of material does she need for the border?

   Answer: *Betsy needs 12 feet of material.*
   Strategy: *Possible Strategy: Draw a Diagram*

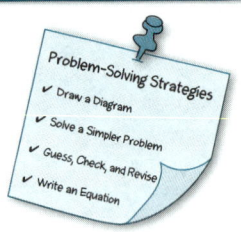

2. The stage for a band concert is shown from above. The middle of the stage is a rectangle that is 60 feet across and 40 feet from back to front. There are three semi-circular sections added to the stage, one on the front, one on the left, and one on the right. To the nearest 10 square feet, what is the area of the stage? Use 3.14 for π.

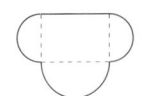

   Answer: *The area of the stage is about 5,070 square feet.*
   Strategy: *Possible Strategy: Solve a Simpler Problem*

3. The formula to convert the number of degrees Fahrenheit to degrees Celsius is
   $$C = \frac{5}{9}(F - 32)$$
   At what temperature is the number of degrees Fahrenheit the same as the number of degrees Celsius? (Hint: It is below zero.)

   Answer: *The number of degrees Celsius is the same as the number of degrees Fahrenheit at −40 degrees.*
   Strategy: *Possible Strategy: Guess, Check, and Revise*

4. Joe is standing next to a flagpole. He casts a shadow 2 feet 6 inches long. The flagpole casts a shadow 10 feet long. Joe is 5 feet tall. How tall is the flagpole?

   Answer: *The flagpole is 20 feet tall.*
   Strategy: *Possible Strategy: Write an Equation*

5. The length and height of a triangular banner are shown. The white triangle in the middle is similar to the banner, with dimensions half those of the entire banner. What is the area of the gray section?

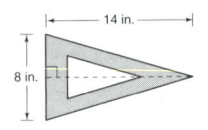

   14 in.
   8 in.

   Answer: *The area of the gray section is 42 square inches.*
   Strategy: *Possible Strategy: Solve a Simpler Problem*

   Explain how you could solve the problem in a different way.

   Sample Answer: The area of the white triangle is $\frac{1}{4}$ the area of the entire banner. The rest of the banner is gray. This means that the area of the gray section is $\frac{3}{4}$ the area of the entire banner, or $\frac{3}{4}$ of 56 square inches.

168 Unit 5 Using Measurement

169

Support for Assessment

The problems on pages 168–171 reflect strategies and mathematics students used in the unit.

Although this unit focuses on four problem-solving strategies, students may use more than one strategy to solve the problems or use strategies different from the focus strategies. Provide additional support for those students who need it.

Draw a Diagram For Problems 1, 6, and 10, ask students how they will use their diagrams to help them solve the problems. For Problem 1, make sure students notice the different units of measure. For Problem 6, ask how the diameter of the semicircular plaza compares to a dimension of the rectangle. For Problem 10, after students have plotted the given points, ask what part of a rectangle they will draw if they connect the points.

Solve a Simpler Problem For Problems 2 and 5, guide students to use the dimensions given to find the dimensions of simpler shapes. Have students explain if they will add or subtract areas of simpler shapes and why.

Guess, Check, and Revise For Problems 3 and 8, ask questions such as, *What measurement will you guess? Why?*

Write an Equation For Problems 4, 7, and 9, encourage students to think about how the information given will determine the order in which they write terms in the equivalent ratios.

You may wish to use the *Review* to assess student progress or as a comprehensive review of the unit.

Promoting 21st Century Skills

Write About It
Communication

When students explain how they made a decision about what problem-solving strategy or strategies to use, they have the opportunity to review the factors and reasoning that led to the decision. Responses should clearly state the choice(s) students made and their reasoning.

Team Project: Design a Sculpture
Collaboration: 3–4 students

Remind students that the goal of the project is to do some group planning, develop ideas individually, and make a group decision based on those ideas. Monitor groups as they work to check that each member is participating in all phases of the project.

Ask questions that help students summarize their thinking. *How did you choose the group's favorite sculpture? What did you consider when deciding where to place the sculpture?*

Extend the Learning
Media Literacy

If you have access to the Internet, navigate to sites where students can access computer drawing programs. They may find the programs helpful in creating diagrams of their sculptures, as well as a sketch of the group's favorite sculpture positioned in the place determined by the group.

🔍 Search **conservation sculpture**

Solve each problem. Show your work. Record the strategy you use.

6. Jon is planning a route for a skating event at Oak Park. The rectangular park is 500 meters × 200 meters, with a semicircular plaza on one of the short sides. The radius of the plaza is 100 meters. To the nearest 10 meters, how long will the route be if it goes completely around the park? Use 3.14 for π.

Answer: The route will be about 1,510 meters long.
Strategy: *Possible Strategy:* Draw a Diagram

7. Olivia found a satellite image of her neighborhood online. She wants to know how far she jogs when she goes once around her block. On the screen, this distance is 30 inches. On the screen, the front of her apartment building is 4 inches wide, but she knows it is actually 220 feet wide. How far is it around Olivia's block?

Answer: It is 1,650 feet around the block.
Strategy: *Possible Strategy:* Write an Equation

8. The length and width of the base of a square prism are the same. The prism's height is 3 times its length. Its volume is 3,993 cubic inches. What are the dimensions of the prism?

Answer: The prism is 11 inches long, 11 inches wide, and 33 inches high.
Strategy: *Possible Strategy:* Guess, Check, and Revise

Explain how your answer fits all the information given in the problem.

Sample Answer: The volume of a prism is found by multiplying the length × width × height. If I multiply the dimensions I found, I get the volume given in the problem.

9. The Great Pyramid of Cholula in Mexico is the world's largest monument. In a picture similar to the actual monument, a side of its base is 9 inches long and its height is $1\frac{5}{16}$ inches. The actual length of a side of the base is 450 meters. To the nearest whole meter, what is its actual height?

Answer: The actual height is about 66 meters.
Strategy: *Possible Strategy:* Write an Equation

10. Ms. Smith wants to put a fence around a rectangular field. She makes a scale drawing on a coordinate grid. Two of the opposite corners are at (15, 45) and (27, 35). The scale is *1 unit = 30 feet*. How many feet of fence does Ms. Smith need to completely enclose the field?

Answer: She needs 1,320 feet of fence.
Strategy: *Possible Strategy:* Draw a Diagram

Write About It
Look back at Problem 6. Explain how you chose a strategy or strategies to solve the problem.

Sample Answer: I read the problem and used the *Draw a Diagram* strategy to visualize the route. Then I used *Write an Equation* to find the answer.

Team Project: Design a Sculpture

Your group is designing a sculpture to represent your school's efforts to conserve resources and help the environment.

Plan
1. Discuss ideas for three different sculptures.
2. Make diagrams of each sculpture. Label the dimensions. Also, label a perimeter or circumference, an area, and/or a volume in each diagram.

Decide As a group, choose your favorite sculpture.

Calculate Find how much space is needed to display the sculpture. Decide where you will place the sculpture, in or outside the school.

Present As a group, share your design with the class. Describe what each sculpture represents. Discuss your reasoning and how you know it will fit in the place you have chosen.

Shapes to Use in the Sculpture
- prisms
- pyramids
- spheres
- cylinders
- cones

UNIT 6: Problem Solving Using Data and Probability

CCSS 8.SP Statistics and Probability

Unit Overview

Lesson	Problem-Solving Strategy	Math Focus
21	Make a Graph	Collect, Organize, Display, and Analyze Data
22	Look for a Pattern	Scatter Plots
23	Make an Organized List	Counting Methods, Permutations, and Combinations
24	Make an Organized List	Probability

Promoting Critical Thinking

Higher order thinking questions occur throughout the unit and are identified by this icon: HOTS. These questions progress through the cognitive processes of remembering, understanding, applying, analyzing, evaluating, and creating to engage students at all levels of critical thinking.

UNIT 6: Problem Solving Using Data and Probability

Unit Theme: Going Places

There are a variety of reasons for where, when, why, and how people travel. Your classmates may walk, bike, take the bus, or ride in a car to get to school. You might know people who use the subway, bus, or train to get to work each day. In this unit, you will see how people use math when they are going places.

Math to Know

In this unit, you will apply these math skills:
- Use graphs to organize, display, and analyze data
- Find combinations and permutations
- Use probability to find the likelihood of events

Problem-Solving Strategies
- Make a Graph
- Look for a Pattern
- Make an Organized List

Link to the Theme

Write another paragraph about the beach activities Michelle wants to try. Include some of the activities from the list.

Michelle's family is at the beach for the day. Her younger brother has made a list of the activities he has planned. He cannot decide which activity to do first. Michelle helps him think of some of the ways he could order them.

Students' paragraphs will vary, but should include some words from the list.

Activities
- Swim
- Walk on the boardwalk
- Build a sandcastle
- Get ice cream
- Explore the tidal pools
- Play beach volleyball

Use Math Language

Review Vocabulary

The list below shows vocabulary terms in this unit. Knowing the meaning of these terms will help you understand the problems.

circle graph   dependent events   line of best fit   probability
combination   independent events   permutation   scatter plot
correlation

Vocabulary Activity — Math Terms

Some terms are found only in math. Use terms from the list above to complete each sentence.

1. I can use a **circle graph** to compare data that represent parts of a whole.
2. A straight line that follows the data points on a scatter plot is called a **line of best fit**.
3. A possible arrangement of a collection of things where order matters is called a **permutation**.
4. I can draw points on a **scatter plot** to show a relationship between two quantities that vary.

Graphic Organizer — Word Web

Complete the graphic organizer.
- In the top oval, write a definition of *probability*.
- In each linked oval, write a different vocabulary term related to *probability*.
- Then write a definition of each related term.

Sample Answers:

probability — the likelihood that an event occurs

dependent events — Two events are dependent events if the occurrence of one of the events affects the probability of the occurrence of the other event

independent events — Two events are independent if the occurrence of neither event affects the probability of the occurrence of the other event

172   173

Link to the Theme — Going Places

Ask students to read the direction line and story starter. If students are having trouble getting started, ask questions such as, *Of the activities listed, which would you like to do first? Which activity would you want to save for last?*

Unit 6 Differentiated Instruction

Extra Support

Some students may need practice drawing angles to make circle graphs in the unit. They may also need reinforcement of concepts in probability.

Angles Demonstrate for students how to use a protractor to draw an angle with a given measurement. Guide students through drawing a 60° angle. Have students draw angles of 80°, 45°, 72°, and 110°. Discuss how drawing a 110° angle is different from drawing the other angles.

Probability Show students a 1-6 number cube. Guide them to find the likelihood, or probability, that the cube will land on an odd number when tossed. Have students make a list of all the numbers on the cube. Explain that these are the possible outcomes. Then have students make a list of all the odd numbers on the cube. Explain that these are the favorable outcomes. Use these two lists to determine that the probability of landing on an odd number is 3 out of 6, or $\frac{1}{2}$. Have students find the probability of other simple events involving a number cube or spinner.

Challenge Early Finishers

Early finishers may enjoy the challenge of identifying missing information in a problem. Provide a problem like the following one which is missing important facts needed to answer the question.

> Alice works at a sandwich shop. The shop offers 3 kinds of meat: ham, turkey, and roast beef. How many different kinds of sandwiches are there with a choice of 1 meat and 1 kind of bread?

Have students consider what additional information is needed to solve the problem and how different possibilities for that information will affect the solution.

English Language Learners

Language Development

Homophones Tell students that some words may sound the same, but have different spellings and different meanings. Write on the board the following sentences from the unit: *What will be the sum of all the percents in the circle graph?* and *Some results are shown in the graph.* Ask a volunteer to circle the words that sound the same, but have different spellings and meanings. Discuss how the words *sum* and *some* are different. Then repeat the activity with this sentence: *The tree diagram should have four levels, one for each day.* Discuss how *four* and *for* are different. Emphasize the importance of reading math problems and lessons carefully to be able to distinguish words that sound the same, but have different meanings.

Vocabulary

Outcome Write the word *outcome* on the board, say it aloud, and have students repeat after you. Explain that an *outcome* is a result of an action. Ask, *What is a possible outcome if I roll this number cube?* Guide students to include the word *outcome* in their responses. For example, students may respond, *Four is a possible outcome.* Then ask, *If I roll this number cube, how many possible outcomes are there?* Guide students to respond that there are six possible outcomes.

Writing

Verb Forms Many of the problems in this unit on data and probability include various forms of the verb *to choose*. Make sure students are able to recognize and use the various forms of this verb. Write the following on the board: *chooses, choose, chose*. Then guide students to use each word once to fill in the following frames:

> Last night, Tina _____ an apple for a snack. Usually, Claudio _____ an orange. Tomorrow, I will _____ a pear.

Point out that these are all various forms of the verb *to choose*.

Lesson 21
Strategy Focus: Make a Graph

Lesson Overview

Lesson Materials: calculator

Skills to Know	Outcome	Math Vocabulary	eResources www.optionspublishing.com
• Use circle graphs to organize, display, and analyze data • Find the percent of a number	Students will recognize that making a graph is an efficient way to solve problems that require analyzing and interpreting data.	circle graph	• Interactive Whiteboard Transparency 21 • Homework, Unit 6 Lesson 21 • Problem-Solving Checklist, also available in the student worktext, page 7

Modeled Instruction

Learn

Ask questions to confirm students' comprehension of the problem's context.

- If the school's suggestion is followed, where will the eighth graders go for their class trip? How do you know?
- What does the destination *Other* represent in the table?

As students read the problem again, pose questions to help them identify important information.

- If Daria's suggestion were followed, what destinations would the students have to choose from?
- How can you find how many students were surveyed in all?

Reread You may wish to use this Think Aloud to demonstrate how to read a problem for different purposes.

I know this problem is about a survey of class trip destinations. Daria thinks another survey should be taken. I need to figure out why she may think this so that I can decide whether it is a good idea.

I will reread the problem and look closely at the survey results. I can see that the numbers of students who voted for New York City, Washington, D.C., Acadia, and Philadelphia are about the same. But are they close enough to each other to make a difference? I wonder how the number of students who voted for New York City compares to the number of eighth graders. I could add and compare the numbers in the table in different ways.

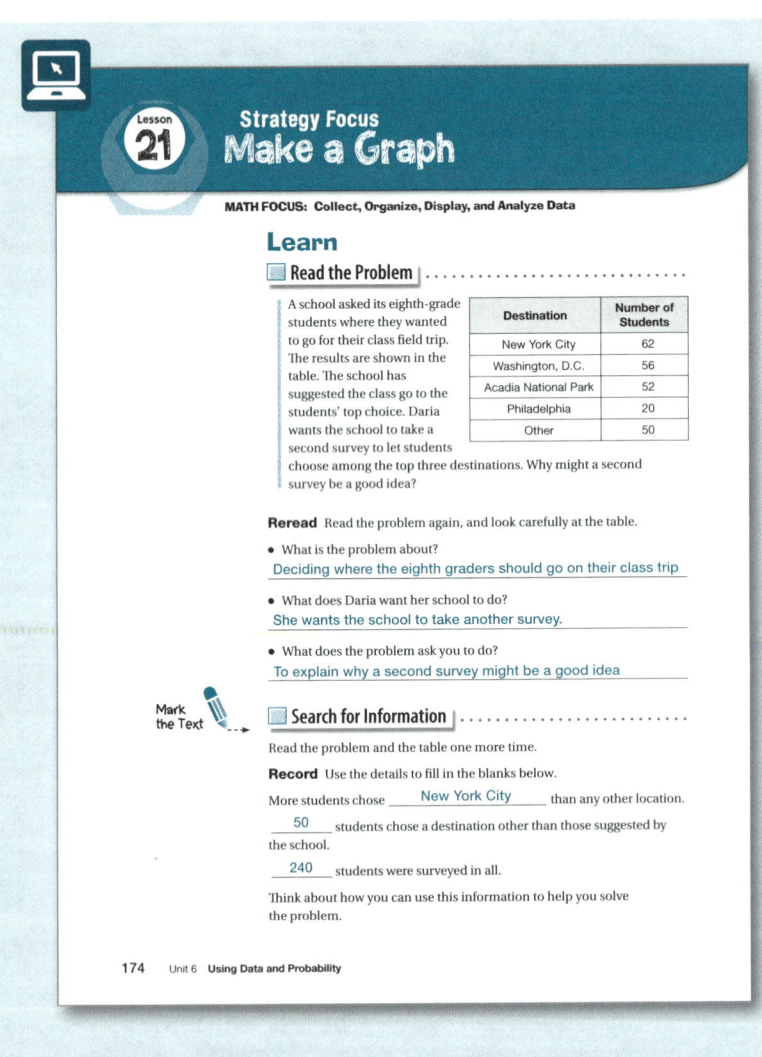

Lesson 21 — Strategy Focus: Make a Graph

MATH FOCUS: Collect, Organize, Display, and Analyze Data

Learn

Read the Problem

A school asked its eighth-grade students where they wanted to go for their class field trip. The results are shown in the table. The school has suggested the class go to the students' top choice. Daria wants the school to take a second survey to let students choose among the top three destinations. Why might a second survey be a good idea?

Destination	Number of Students
New York City	62
Washington, D.C.	56
Acadia National Park	52
Philadelphia	20
Other	50

Reread Read the problem again, and look carefully at the table.

- What is the problem about?
  Deciding where the eighth graders should go on their class trip
- What does Daria want her school to do?
  She wants the school to take another survey.
- What does the problem ask you to do?
  To explain why a second survey might be a good idea

Search for Information

Read the problem and the table one more time.

Record Use the details to fill in the blanks below.

More students chose ___New York City___ than any other location.

___50___ students chose a destination other than those suggested by the school.

___240___ students were surveyed in all.

Think about how you can use this information to help you solve the problem.

174   Unit 6 Using Data and Probability

I wonder if making some type of visual display of the data would make it easier for me to analyze it and evaluate whether Daria's idea is a good one.

114   Unit 6

Decide What to Do

You know the destinations and how many students chose each one.

Ask How can I explain why the school might want to consider a second survey?

- I can find the fraction of students who chose each destination.
- I can *Make a Graph* using the fractions to answer the question.

Use Your Ideas

Step 1 Write a fraction that shows the relationship between the number of students who chose each destination and the total number of students.

Step 2 To make a circle graph, multiply each fraction by 360° to find the measure of the central angle for that part.

Destinations	Students	Fractions	Angles
New York City	62	$\frac{62}{240}$	93°
Washington, D.C.	56	$\frac{56}{240}$	84°
Acadia National Park	52	$\frac{52}{240}$	78°
Philadelphia	20	$\frac{20}{240}$	30°
Other	50	$\frac{50}{240}$	75°

Step 3 Make and label a circle graph. Study your circle graph.

There are two reasons that a second survey might be a good idea.

1. The graph shows that almost $\frac{3}{4}$ of the students did not pick the top choice.
2. The circle graph shows that more than $\frac{1}{4}$ of the students did not pick one of the top three choices.

Use approximate fractions to complete the sentences.

Review Your Work

Does your circle graph accurately represent the data?

Visualize How did the circle graph help you analyze the problem?

Sample Answer: I could relate parts of the circle to each other visually much quicker than I could add the fractions.

175

Modeled Instruction *(continued)*

Help students make a connection between the facts they know and using a graph to solve the problem.

- *What types of graphs might you use to display the data?*

Pose questions that give meaning to each step in the solution process.

- *Why is a circle graph a good choice for showing the data?*
- *Why do you multiply the fraction for each destination by 360°?*

Emphasize to students the importance of checking that the graph accurately represents data.

- *How can you check that your graph is an accurate representation of the data?*

HOTS Visualize Responses might indicate that it is quicker to visually recognize the relationships of parts to the whole than to add and compare fractions.

Guided Practice

Try

① Prompt students to consider how the circle graph can be used to solve the problem.

Discuss the context of the problem as well as the information needed to solve it.

- *What do the numbers in the circle graph show?*

Ask students to explain how they used the percents to find the actual numbers.

- *What equations did you use to find how many families chose amusement parks? The beach?*

Encourage students to use the circle graph to check that their answers are reasonable.

HOTS Describe Responses should show that students recognize that Kathy subtracted percents rather than numbers of families.

Try

Solve the problem.

① Two hundred families answered a survey on what activities they would look forward to most during a trip to Florida. How many more families chose amusement parks than chose the beach?

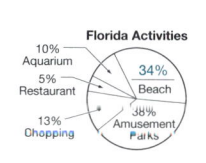

Florida Activities
- 10% Aquarium
- 5% Restaurant
- 34% Beach
- 13% Shopping
- 38% Amusement Parks

Mark the Text

Ask Yourself: What will be the sum of all the percents in the circle graph?

Read the Problem and Search for Information

Locate the important details in the problem and in the circle graph. Reread and underline the question you will answer.

Decide What to Do and Use Your Ideas

You need to find the missing value in the circle graph and use it to answer the question.

Step 1 All the families interviewed are represented in the circle graph. Fill in the missing percent.

$100\% - (38\% + 13\% + 5\% + 10\%) = \underline{34}\%$

Step 2 Use the information in the problem and the appropriate percentages from the circle graph to complete the sentences.

- $\underline{200}$ families were surveyed.
- $\underline{76}$ families chose amusement parks in the survey.
- $\underline{68}$ families chose the beach in the survey.

So $\underline{8}$ more families chose amusement parks than the beach.

Review Your Work

Does your answer make sense when you compare the two sections of the circle graph?

Describe Kathy said that 4 more families chose amusement parks than chose the beach. What error might she have made?

Sample Answer: Kathy might have subtracted the percents instead of subtracting the number of families.

176 Unit 6 Using Data and Probability

Lesson 21 115

Scaffolded Practice
Apply

**② ** Make sure students understand how the graph relates to the number of parking spaces needed.
- *Which part of the circle graph relates to the number of parking spaces the company needs?*
- *How can you use the graph to find the number of people who need parking spaces?*

HOTS Interpret Responses should demonstrate that students understand how to find the percent of a whole.

**③ ** Help students recognize that the numbers on the graphs show the number of states the students in a given grade have lived in. Ask students how they can use the graphs to solve the problem.
- *To solve this problem, do you need to know the actual percents of the second- and eighth-graders represented by the parts of the graphs? Explain.*

HOTS Identify Responses should show that students understand how to rewrite a fraction as a percent.

**④ ** Ask students to identify the steps they will use to solve the problem.
- *How will you determine the percent of students who chose Other?*
- *How will you determine the number of students who chose Other?*

HOTS Choose Responses should demonstrate students' understanding that a circle graph shows how data represent the parts of a whole and that percents are parts of a whole.

**⑤ ** Make sure students recognize that the graph represents only the first half of the trip.
- *What do the percents in the graph tell you?*
- *How do the percents relate to the whole trip?*

HOTS Model Explanations should describe how students could represent each half of the trip as half of a circle graph. Students may note that they do not have the data to complete the graph for the second half of the trip.

116 Unit 6

Apply
Solve the problems.

② A company asked its employees how they commute to work. Some results are shown in the graph. There are 1,200 employees. How many parking spaces does the company need?

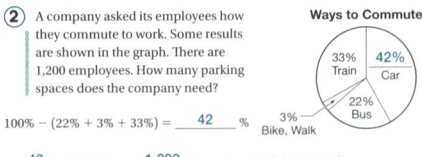

Ways to Commute: 42% Car, 33% Train, 22% Bus, 3% Bike, Walk

$100\% - (22\% + 3\% + 33\%) = \underline{42}\%$

Ask Yourself How can I find the percent of people who drive a car to work?

$\underline{42}\%$ of the $\underline{1{,}200}$ employees drive to work.

Answer The company needs at least 504 parking spaces.

Hint To change a percent to a number of people, you need to use the total number of people in the survey.

Interpret How did you know what operations you needed to use?
Sample Answer: I knew to multiply the percent by the number of employees because a percent of a whole means to multiply.

③ Students in second and eighth grade were asked how many states they had lived in. Which class has a greater percent of students who have lived in three or more states?

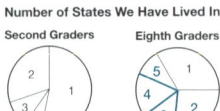

Number of States We Have Lived In
Second Graders: 2, 1, 3, 4
Eighth Graders: 5, 1, 4, 3, 2

Hint Use the incomplete graph to show the data for the eighth-grade students.

Number of States	Number of 8th Graders	Fraction of Whole	Degrees in Circle Graph
1	10	$\frac{10}{30}$	120°
2	8	$\frac{8}{30}$	96°
3	4	$\frac{4}{30}$	48°
4	5	$\frac{5}{30}$	60°
5	3	$\frac{3}{30}$	36°

Ask Yourself Should I include the number of students who have lived in three states to find the answer?

Answer The eighth-grade class has a greater percent of students who have lived in three or more states.

Identify What fraction of the eighth-grade students have lived in 5 different states? What percent of the students is that?
Sample Answer: $\frac{3}{30}$, or $\frac{1}{10}$, of the eighth-grade students have lived in 5 different states. That is 10% of the class.

Lesson 21 Strategy Focus: Make a Graph 177

---

④ Omar started to make a circle graph about travel. He planned to finish it at home, but left his data at school. Omar recalls that there were 36 people surveyed and that twice as many students chose "boat" as chose "other." How many students chose "other"?

How We Like to Travel: 41% Plane, 28% Boat, 14% Other, 11% Train, 6% Car

Ask Yourself How can I find the total percent for the two unlabeled parts of the circle graph?

$\underline{42}\%$ of students chose "boat" or "other."

$\underline{14}\%$ of students chose "other."

Hint Remember that you cannot have a fraction of a student.

Answer 5 students chose "other" in Omar's survey.

Choose Why is a circle graph a good way to show this data?
Sample Answer: A circle graph is a good way to show data that represent parts of a whole, and percents are parts of a whole.

⑤ Five people shared the driving on a 750-mile trip. Halfway through the trip, Mark made the circle graph shown. He said, "I have already driven more than my 20% share." Is Mark right? Explain.

Hint Remember that the circle graph is made before the trip is complete.

20% of the total trip is $\underline{150}$ miles.

Mark has driven $\underline{131.25}$ miles.

Fair Shares?: 35% Mark, 20% Theo, 18% Cleo, 15% Hal, 12% May

Ask Yourself How can I find the total number of miles that Mark has driven?

Answer No, Mark should still drive 18.75 miles.

Model How could you change the circle graph in this problem to represent the total driving distance?

Sample Answer: The circle would be marked for the whole trip. Each person's angle and percentage would be half as big. The rest would represent the part that had not yet been driven.

178 Unit 6 Using Data and Probability

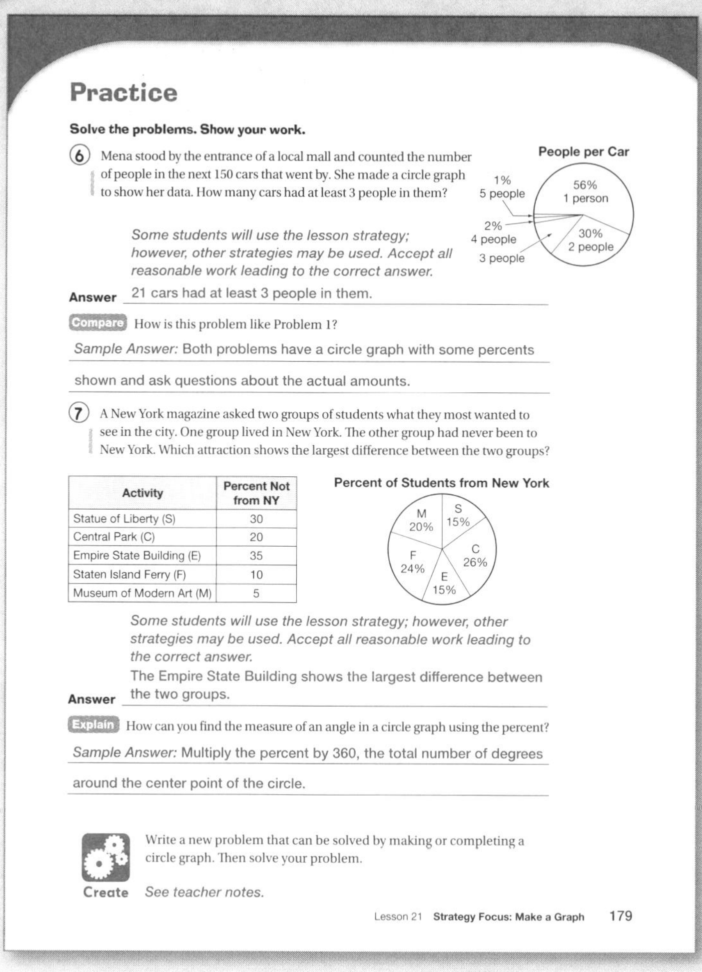

Practice

Solve the problems. Show your work.

6 Mena stood by the entrance of a local mall and counted the number of people in the next 150 cars that went by. She made a circle graph to show her data. How many cars had at least 3 people in them?

People per Car
- 1% 5 people
- 56% 1 person
- 2% 4 people
- 30% 2 people
- 3 people

Some students will use the lesson strategy; however, other strategies may be used. Accept all reasonable work leading to the correct answer.

Answer 21 cars had at least 3 people in them.

Compare How is this problem like Problem 1?

Sample Answer: Both problems have a circle graph with some percents shown and ask questions about the actual amounts.

7 A New York magazine asked two groups of students what they most wanted to see in the city. One group lived in New York. The other group had never been to New York. Which attraction shows the largest difference between the two groups?

Activity	Percent Not from NY
Statue of Liberty (S)	30
Central Park (C)	20
Empire State Building (E)	35
Staten Island Ferry (F)	10
Museum of Modern Art (M)	5

Percent of Students from New York
- M 20%
- S 15%
- C 26%
- F 24%
- E 15%

Some students will use the lesson strategy; however, other strategies may be used. Accept all reasonable work leading to the correct answer.

Answer The Empire State Building shows the largest difference between the two groups.

Explain How can you find the measure of an angle in a circle graph using the percent?

Sample Answer: Multiply the percent by 360, the total number of degrees around the center point of the circle.

Create Write a new problem that can be solved by making or completing a circle graph. Then solve your problem.
See teacher notes.

Lesson 21 Strategy Focus: Make a Graph 179

Create

In this lesson, students write their own problem that can be solved using a circle graph. If students are struggling, suggest they make a circle graph showing percents with one percent missing, as in Problem 2. Then they can ask a question about finding an actual number.

Accept responses that include a problem in which students make a circle graph to show how data representing parts are related to the whole, or in which students find the data missing from a circle graph. Responses should include a correct solution.

Independent Practice
Practice

Students should be encouraged to choose any strategy to solve Problems 6 and 7, though many may prefer to *Make a Graph*.

6 Some students may subtract the percent of cars that had less than 3 people in them from 100% to find the percent that had at least 3 people in them.

Sample Work

Percent of cars with 3 people in them:

$100\% - (2\% + 1\% + 56\% + 30\%) = 11\%$

Percent of cars with at least 3 people in them:

$11\% + 2\% + 1\% = 14\%$

Number of cars with 3 or more people in them:

$0.14 \times 150 = 21$

Compare Responses should demonstrate that students recognize that for both problems, they must find actual numbers when given percents.

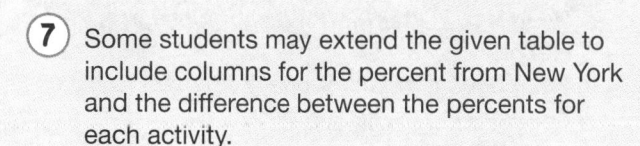

7 Some students may extend the given table to include columns for the percent from New York and the difference between the percents for each activity.

Sample Work

Activity	Percent Not From NY	Angle
S	30	108°
C	20	72°
E	35	126°
F	10	36°
M	5	18°

Percent of Students Not From New York 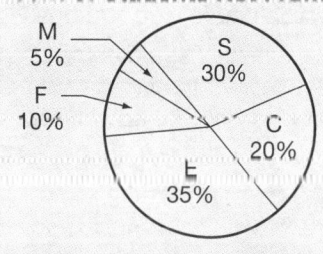
- M 5%
- S 30%
- F 10%
- C 20%
- E 35%

Explain Students' explanations should show an understanding that the measure of an angle in a circle graph is a percent of 360°, the total number of degrees in a circle.

Lesson 21 117

Lesson 22: Strategy Focus — Look for a Pattern

Lesson Overview

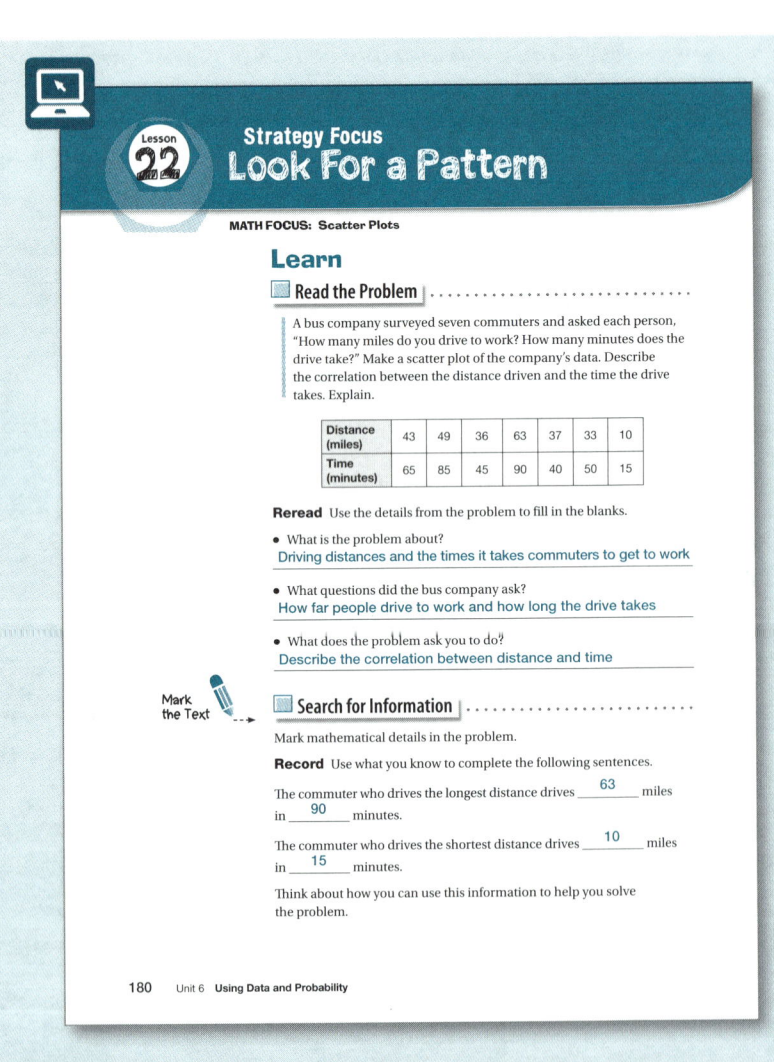

Lesson Materials: ruler

Skills to Know	Outcome	Math Vocabulary	eResources www.optionspublishing.com
• Make scatter plots to organize, display, and analyze data • Plot ordered pairs on a coordinate grid	Students will recognize that looking for a pattern in a scatter plot is an efficient way to solve problems that involve two sets of data.	correlation, line of best fit, scatter plot	• Interactive Whiteboard Transparency 22 • Homework, Unit 6 Lesson 22 • Problem-Solving Checklist, also available in the student worktext, page 7

Modeled Instruction

Learn

To be sure students understand the context of the problem, ask questions like the ones below.

- *What is a scatter plot? When might you use one?*
- *What does it mean for one set of data to correlate to another set of data?*

As students read the problem again, ask questions to help them focus on the details needed to solve it.

- *What do the data in each column of the table represent?*
- *Are the data in the table in any particular order? Explain.*

Reread You may wish to use this Think Aloud to demonstrate how to read a problem to decide on a problem-solving strategy.

This problem is about a survey of some commuters. Two sets of data were collected: the number of miles the commuters drive to work and the number of minutes their drives take. I am asked to find if there is a relationship between these two sets of data.

I will reread the problem and look closely at the survey results. Each column gives the miles and minutes for one commuter. But neither the miles nor the minutes are arranged in order from least to greatest or greatest to least. So I cannot see any relationship between the sets of data in the table.

Maybe I could rewrite the table so that the miles are listed from least to greatest. But what happens if the minutes do not turn out to continually increase or continually decrease? The problem suggests I make a scatter plot of the data. I wonder if it will show a relationship between the miles and minutes for each commuter.

Lesson 22 — Strategy Focus: Look For a Pattern

MATH FOCUS: Scatter Plots

Learn

Read the Problem

A bus company surveyed seven commuters and asked each person, "How many miles do you drive to work? How many minutes does the drive take?" Make a scatter plot of the company's data. Describe the correlation between the distance driven and the time the drive takes. Explain.

Distance (miles)	43	49	36	63	37	33	10
Time (minutes)	65	85	45	90	40	50	15

Reread Use the details from the problem to fill in the blanks.

- What is the problem about?
  Driving distances and the times it takes commuters to get to work
- What questions did the bus company ask?
  How far people drive to work and how long the drive takes
- What does the problem ask you to do?
  Describe the correlation between distance and time

Mark the Text

Search for Information

Mark mathematical details in the problem.

Record Use what you know to complete the following sentences.

The commuter who drives the longest distance drives __63__ miles in __90__ minutes.

The commuter who drives the shortest distance drives __10__ miles in __15__ minutes.

Think about how you can use this information to help you solve the problem.

180 Unit 6 Using Data and Probability

---

118 Unit 6

Decide What to Do

You know the times and distances of the commutes for 7 drivers.

Ask How can I describe the correlation between distance and time?

- I can make a scatter plot and use the strategy *Look for a Pattern* to see if the data seem to follow a pattern.
- I can draw a line of best fit through the points to determine the type of correlation between the distance driven and the time the drive takes.

Use Your Ideas

Step 1 Make a scatter plot for this set of data.

Step 2 Draw a line of best fit through the data points.

Step 3 Determine the type of correlation.

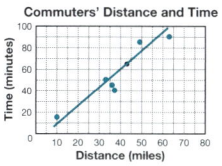

Commuters' Distance and Time

Draw a line of best fit so that it goes as close as possible to as many of the points as possible.

- Is the slope of the line positive, negative, or zero? __Positive__
- Are all or most of the points close to the line? __Yes__

The number of miles driven has a __positive__ correlation with the number of minutes driven because the data points are all __close to a line with a positive slope__.

Review Your Work

Does your scatter plot show the data accurately?

Generalize Why is it important to pay attention to the sample size when you make a conclusion from a set of data?

Sample Answer: A pattern in 7 points may not hold true for a greater number of points.

Modeled Instruction (continued)

Help students make a connection between the facts they know and a strategy they can use.

- *How might a scatter plot show whether or not there is a relationship between two sets of data?*

Pose questions that help students focus on the steps used to solve the problem.

- *How does the line of best fit compare to the location of the points on the scatter plot?*
- *What does a positive correlation tell you about how the miles and minutes driven are related?*

Emphasize the importance of checking that the data are plotted accurately in a scatter plot.

- *How could plotting a data point incorrectly affect the usefulness of a scatter plot?*

HOTS Generalize Responses should explain that the line of best fit approximates a pattern between two data sets, and the larger the sample size, the better the approximation.

Try

Solve the problem.

1. A commuter rail company counted people leaving the station each hour from 6 A.M. to 9 P.M. for 3 days. One train car seats 40 passengers. For which hours should the train service schedule more than 5 cars?

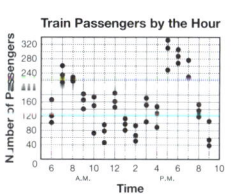

Train Passengers by the Hour

Read the Problem and Search for Information

Underline details that will help you solve the problem.

Decide What to Do and Use Your Ideas

I need to look for a pattern in the data that is true for all three days.

Step 1 If the company schedules more than 5 cars on a train, how many passengers are they expecting? __More than 200__

Step 2 Draw a line on the graph that will help you see which hours have enough passengers to require more than 5 cars.

The train service should schedule more than 5 cars for these hours: __7 A.M., 8 A.M., 5 P.M., 6 P.M., and 7 P.M.__

Ask Yourself: Should I draw a horizontal line or a vertical line to show a particular number of passengers on this graph?

Review Your Work

Does the line you drew on the graph separate the points that require more than 5 train cars from those that do not?

Conclude If the trains at the busiest hours have 6 cars, will there ever be passengers who cannot find a seat at those hours? Explain.

Sample Answer: Yes, even 6 cars may not be enough for many trains, and there could always be more passengers than the company expects.

Guided Practice

Try

1. Prompt students to consider how a scatter plot is useful for analyzing a set of data, and how students can use one to solve the problem.

    Discuss any patterns students may notice on the scatter plot within the context of the problem.

    - *Why do you think the numbers of passengers increase or decrease during certain periods of time?*

    Have students explain how they decided where to draw the line on their scatter plots.

    - *Where will you draw your line on the scatter plot? Why?*

    Ask students to explain how they know their line is drawn so that the times when more than 5 cars are needed are identified.

    HOTS Conclude Students' responses should show an understanding that 6 cars will have seats for 240 passengers, and there are times when the trains will have more than 240 passengers.

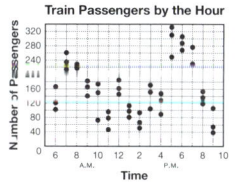

Lesson 22   119

Scaffolded Practice
Apply

② Make sure students understand the context of the problem and the question it asks.
- *What are the two parts of the park's policy for the number of employees?*
- *How can you find the greatest number of visitors allowed for 60 employees?*

HOTS Describe Responses should note how the patterns in the number of visitors increases and decreases over the course of the day.

③ The lack of correlation in this problem is contrary to the logical assumption that customers who shop longer will spend more. Prompt students to explain why it is important to have many data points to confirm a correlation.
- *Why was it better to survey 12 customers instead of 6 or fewer customers?*

HOTS Demonstrate Explanations should note that an increase in the values in one set of data and a decrease in the values in the other set of data result in a negative correlation.

④ Guide students to recognize that the line of best fit in this problem indicates no correlation between the data sets. To help students solve this problem, ask questions like this.
- *How does the scatter plot help you analyze the data more efficiently than using a table?*

HOTS Determine Students' explanations should indicate that a horizontal line of best fit can help them give an approximate average.

⑤ Encourage students to describe a pattern in the data. To help students solve this problem, ask questions like this.
- *Will a straight line on the scatter plot help you see a pattern in the data? Explain.*

HOTS Interpret Responses should show that students understand how the data in this problem are related and that they can identify a relationship between temperature and number of trips.

120 Unit 6

Apply
Solve the problems.

② The graph shows attendance at an amusement park. The park must have 1 employee present for every 50 visitors, but never fewer than 60 employees present. Based on the graph, during which hours can the park use only a minimum staff of 60 employees?

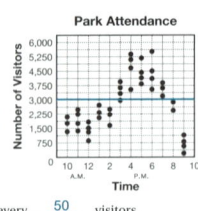

Park Attendance

Ask Yourself What is the greatest number of expected visitors for which the park can have a minimum staff?

The park needs at least 1 employee for every __50__ visitors.
Answer The park can use a minimum staff from 10 A.M. to 2 P.M. and from 8 P.M. to 9 P.M.

Hint Draw a line on the graph that shows the maximum attendance that will allow a minimum staff.

Describe What patterns do you see in the graph?
Sample Answer: Attendance starts out at around 1,500 visitors and increases until about 6 P.M. Then it decreases rapidly.

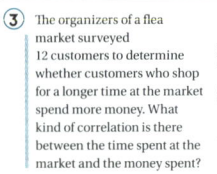

③ The organizers of a flea market surveyed 12 customers to determine whether customers who shop for a longer time at the market spend more money. What kind of correlation is there between the time spent at the market and the money spent?

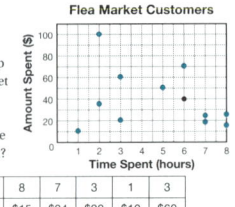

Flea Market Customers

Hint If there is a clear trend in the data, the points will lie approximately along a line or a curved path.

Time Spent (hours)	6	8	7	3	1	3
Amount Spent ($)	$40	$15	$24	$20	$10	$60
Time Spent (hours)	2	5	6	7	8	2
Amount Spent ($)	$35	$50	$70	$18	$25	$100

Answer There is no correlation between the time spent at the market and the amount spent.

Ask Yourself What would the graph look like if customers who spent more time at the market tended to spend more money?

Demonstrate Pick four data points and explain why they show that people who spend more time at the market spend less money.
Sample Answer: (2, 100); (3, 60); (5, 50); (6, 40); (7, 24); and (8, 15) have a negative correlation.

Lesson 22 Strategy Focus: Look For a Pattern 183

Ask Yourself Both axes have about the same scale. How can I be sure that I am plotting the points correctly?

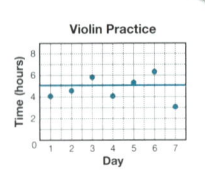

Hint You can look for the greatest increase in either the graph or the table, whichever is easier for you.

④ Trina keeps track of the number of hours she practices each day at music camp. Make a scatter plot. Describe the correlation between the day at camp and the number of hours Trina practices.

Violin Practice

| Day | 1 | 2 | 3 | 4 | 5 | 6 | 7 |
| Time (hours) | 4 | 4.5 | 5.75 | 4 | 5.25 | 6.25 | 3 |

The slope of the line of best fit is about __0__.

Answer There is no correlation between Trina's practice time and the time spent at camp.

Determine Can you use the scatter plot and line of best fit to find an average number of hours that Trina practices each day at camp? Explain.
Sample Answer: Yes, you can see that the daily times are mostly just above and below the line of best fit at five hours.

Hint You may want to draw a line that divides the graph into destinations with more than 5 trips and those with fewer than 5 trips.

⑤ A travel agency records the mean high temperature for a destination whenever a trip is booked. Use the graph of the data at the right. If a destination has a mean high temperature of 60°F in December, would you expect the travel agency to book more than 5 trips or fewer than 5 trips to that place? Explain.

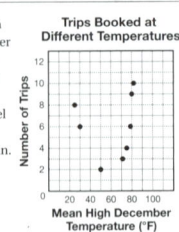

Trips Booked at Different Temperatures

Ask Yourself It looks as though points on this graph may follow a curve. How will that affect using a line of best fit?

Would you expect 60°F to fall above or below the line suggested by the hint? __Below__

Answer I would expect fewer than 5 trips to be booked. The graph shows fewer than 3 trips booked for such a destination.

Interpret Read the graph and describe the pattern in the data.
Sample Answer: In December, more travelers go to destinations that are cold or hot than go to places with moderate temperatures.

184 Unit 6 Using Data and Probability

Independent Practice

Solve the problems. Show your work.

6 Jared went on a camping trip with his scout troop. He recorded the distance the troop hiked each day. Make a scatter plot of Jared's data. Describe the type of correlation between the number of the day and the distance hiked.

Day	1	2	3	4	5	6	7
Distance (miles)	11.25	10.75	9.5	3	8	6.5	4

Some students will use the lesson strategy; however, other strategies may be used. Accept all reasonable work leading to the correct answer.

Answer There is a negative correlation between the number of the day and the distance hiked.

Analyze Which point in Jared's data seems most out of place? Explain.

Sample Answer: The point (4, 3) seems out of place because it is the farthest away from the line of best fit.

7 Kyle is on a cycling vacation. Each day, he records the total number of miles he has biked. The graph shows the first 5 days of his data. If he continues at his current pace, do you expect that Kyle will bring his total up to 100 miles by Day 7? Explain.

Some students will use the lesson strategy; however, other strategies may be used. Accept all reasonable work leading to the correct answer.

Answer Yes, because a line of best fit shows more than 100 miles by Day 7.

Explain How does looking for a pattern help you solve this problem?

Sample Answer: If I find a pattern by drawing a line of best fit, I can predict how many miles Kyle will bike by Day 7.

Create Make up your own situation for a scatter plot. Write a new problem that can be solved using a pattern in your data, then solve it.

See teacher notes.

Lesson 22 Strategy Focus: Look For a Pattern 185

Create

In this lesson, students write their own problem involving analyzing a scatter plot. If students are struggling with writing a new problem, suggest they review the problems in the lesson for ideas.

Accept student responses that include a problem about two related sets of data. The data may be given in a table or in a scatter plot. The problem should ask a question about the relationship between two sets of data. Responses should include a correct solution.

Independent Practice
Practice

Students should be encouraged to choose any strategy to solve Problems 6 and 7, though many may prefer to *Look for a Pattern*.

6 Some students may be able to describe the correlation by simply studying the data in the table.

Sample Work

Jared's Camping Trip

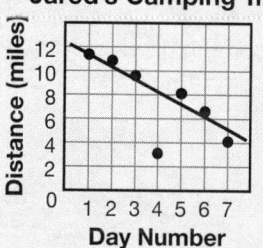

HOTS Analyze Responses should show that students can use a line of best fit on a scatter plot to identify data that do not follow the same pattern as the rest of the data.

7 Ask students if they will draw a straight or a curved line of best fit. Students who draw a curved line to accommodate the data on Days 4 and 5 may conclude that Kyle will not bike 100 miles by Day 7.

Sample Work

Kyle's Miles

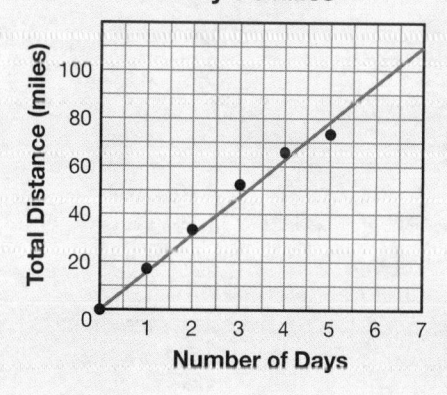

HOTS Explain Students' explanations should demonstrate understanding that a line of best fit with a positive slope indicates that the data for this problem show an increase over time.

Lesson 22   121

Lesson 23
Strategy Focus: Make an Organized List

Lesson Overview

Skills to Know	Outcome	Math Vocabulary	eResources www.optionspublishing.com
• Find combinations and permutations	Students will recognize that making an organized list is an efficient way to find all the possible combinations or permutations for a given situation.	combination, permutation	• Interactive Whiteboard Transparency 23 • Homework, Unit 6 Lesson 23 • Problem-Solving Checklist, also available in the student worktext, page 7

Modeled Instruction
Learn

Probe students' understanding of the problem's context by asking questions similar to the following.

- *How are each of Hannah's caps different?*
- *What is an example of one way Hannah will wear her caps on the four days? What is an example of how she will* not *wear the caps?*

As students read the problem again, guide them to identify the words and numbers needed to solve the problem.

- *How many pink caps does Hannah have? How do you know?*
- *How many days will Hannah wear her pink cap? Why?*

Reread You may wish to use this Think Aloud to demonstrate how to read a problem to determine a problem-solving strategy.

This problem is about Hannah wearing different caps. The problem asks me to find how many ways there are for Hannah to wear her caps during her vacation.

I am going to reread to see what information I can use to solve the problem. I am told that Hannah has four caps that are all different colors. I also see that her vacation is four days long. The problem tells me that Hannah wants to wear a different cap each day of her vacation. That makes me think that Hannah will wear each cap only once during her trip.

To wear the caps in different ways, Hannah could change the order of the colored caps she wears. For example, she could wear pink one day, then white the next day, then blue, then tan. A different order could be blue, then pink, then tan, then white. I bet there are a lot of different orders. I need to organize my work so I do not repeat or leave out any combinations.

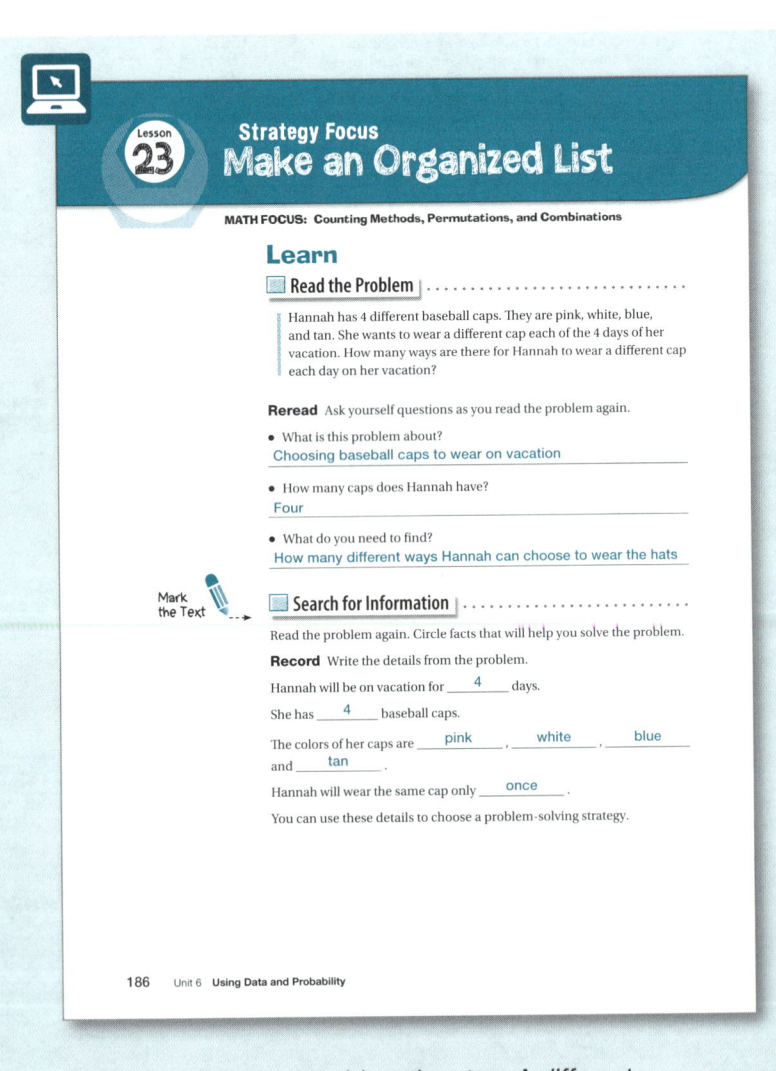

Lesson 23 — Strategy Focus: Make an Organized List

MATH FOCUS: Counting Methods, Permutations, and Combinations

Learn
Read the Problem

Hannah has 4 different baseball caps. They are pink, white, blue, and tan. She wants to wear a different cap each of the 4 days of her vacation. How many ways are there for Hannah to wear a different cap each day on her vacation?

Reread Ask yourself questions as you read the problem again.

- What is this problem about?
  Choosing baseball caps to wear on vacation
- How many caps does Hannah have?
  Four
- What do you need to find?
  How many different ways Hannah can choose to wear the hats

Search for Information

Mark the Text

Read the problem again. Circle facts that will help you solve the problem.

Record Write the details from the problem.

Hannah will be on vacation for __4__ days.

She has __4__ baseball caps.

The colors of her caps are __pink__, __white__, __blue__, and __tan__.

Hannah will wear the same cap only __once__.

You can use these details to choose a problem-solving strategy.

186 Unit 6 Using Data and Probability

---

122 Unit 6

Decide What to Do

You know the number and colors of the different caps Hannah has to choose from. You know the number of days she is on vacation.

Ask How can I find the number of ways she can wear a different hat each day?

- I can use the strategy *Make an Organized List* to show all the permutations of the four hats.

Use Your Ideas

Step 1 Write the choices of caps for the first day in the first row.

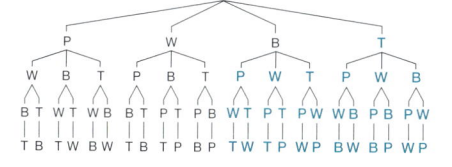

```
            P              W              B              T
         / | \          / | \          / | \          / | \
        W  B  T        P  B  T        P  W  T        P  W  B
       /|  /|  /|     /|  /|  /|     /|  /|  /|     /|  /|  /|
      B T W T W B    B T P T P B    W T P T P W    W B P B P W
      T B T W B W    T B T P B P    T W T P W P    B W B P W P
```

Use P for pink, W for white, B for blue, and T for tan. When you are counting Hannah's different choices you will need to be sure that you count every possibility and that you do not count any possibility more than once.

Step 2 Write the caps that are left for the second day in the second row. Remember, Hannah will wear a given cap only once.

Step 3 Complete the tree diagram for the third and fourth days. Count the branches at the end of the tree.

There are __24__ ways for Hannah to wear a different cap each day on her vacation.

Review Your Work

Look back at your completed tree diagram. Make sure you did not list the same cap more than once in any branch of the tree diagram.

Describe How did the tree diagram help you solve the problem?

Sample Answer: The organized list helped me to be sure that I did not miss or repeat any choices.

187

Try

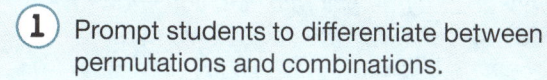

Solve the problem.

1. Alice, Ben, Cara, and Dave are on a photography trip with Mrs. Martinez. In the morning, Mrs. Martinez chooses two people to carry her camera equipment. In the afternoon, the two other people carry it back. How many ways can Mrs. Martinez choose two people to carry the equipment in the morning?

Mark the Text

Read the Problem and Search for Information

Reread the problem. Underline the question you need to answer.

Decide What to Do and Use Your Ideas

You can *Make an Organized List* to find the different ways to choose two people. Then you can count the combinations.

Ask Yourself
If I list Alice and Ben, do I need to list Ben and Alice?

Step 1 Make a table of the different ways Mrs. Martinez can make her choices for the morning. Start by listing the ways if Alice is one of the choices.

First Choice	Second Choice
Alice	Ben
Alice	Cara
Alice	Dave
Ben	Cara
Ben	Dave
Cara	Dave

Step 2 Count the choices in the table.

So there are __6__ different ways Mrs. Martinez can choose two people to carry her equipment in the morning.

Review Your Work

Make sure your table lists every combination of two people.

Conclude A student says that the number of ways to choose two people for the morning and two for the afternoon must be twice the number of ways to choose just the morning. Is this true? Explain.

Sample Answer: No, because when two people are chosen for the morning, the people who carry the equipment in the afternoon have been determined. There are still only 6 ways to choose for both the morning and the afternoon.

188 Unit 6 Using Data and Probability

Scaffolded Practice
Apply

② Make sure students recognize that Jamal's family will not visit all of the museums.
- Why are there only three levels of choices in the diagram when there are five choices of museums?

HOTS Identify Responses should show an understanding that *different* implies that the same museum cannot be chosen twice in the same permutation.

③ Help students understand that an activity can be repeated and that the order of the activities matters.
- Once a morning activity has been chosen, how many choices are there for an afternoon activity? How do you know?

HOTS Contrast Responses should show that students recognize when an item can be used more than once and when order matters.

④ Ask students to explain how they can use a tree diagram to solve this problem.
- How does a tree diagram help you know you have found all the choices without repeating or omitting any?

HOTS Apply Responses should demonstrate students' understanding of the information given in the problem and what questions can be answered using that information.

⑤ Prompt students to explain how they can use an organized list to solve this problem.
- How were the permutations in the first three lines of the list organized? In the last line?

HOTS Relate Responses should show that choosing a different sandwich every day reduces the number of choices.

Apply

Solve the problems.

② Jamal's family has time to go to three different museums while on vacation. Their choices are the air and space museum, the car museum, the modern art museum, the natural history museum, and the science museum. How many ways can they go to three museums if they go to the science museum first?

```
           S
    ┌──┬───┴───┬──┐
    A    C     M   N
   ┌┴┐  ┌┴┐   ┌┴┐ ┌┴┐
   C M N A M N A C N A C M
```

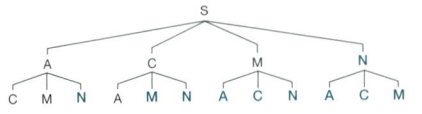

Hint Start with the science museum. Then list all the possibilities for the second and third museum.

Answer There are 12 ways they can go to three museums if they go to the science museum first.

Ask Yourself If Jamal's family knows they will go to the science museum first, how many choices are there for the second museum?

Identify What word in the problem told you that Jamal's family would not go to the same museum twice?

Sample Answer: The word *different* told me the family would not go to the same museum twice.

③ Dante can choose one morning activity and one afternoon activity each day at camp. He can choose arts and crafts, swimming, hiking, or canoeing, and he can choose the same activity twice. How many different ways can Dante make his choices?

Hint You don't always need to make a table to show the possibilities. Start by listing all the choices if he does arts and crafts in the morning.

List all the possible ways to do two activities.

AA, AS, AH, AC
SA, SS, SH, S __C__
HA, HS, H __H__ , HC
CA, __CS__ , __CH__ , CC

Answer Dante has 16 different choices.

Ask Yourself Is "arts and crafts, then swimming" different than "swimming, then arts and crafts"?

Contrast In this problem and in Problem 1, you needed to choose two things from a list of four. How are the two problems different?

Sample Answer: In Problem 1, you cannot choose the same thing both times. The order also did not matter in Problem 1, but it does in this problem.

Lesson 23 Strategy Focus: Make an Organized List 189

---

④ Kelsey is choosing where to go for breakfast, lunch, and dinner while on vacation. Her breakfast choices are Jim's Place or Waffle Land. For lunch, she can go to Burger Palace, Souper Lunch, or Louie's Diner. For dinner, she can go to All-Star Buffet, Pasta Please, or Seafood Bonanza. How many ways can Kelsey choose a place for breakfast, lunch, and dinner?

Hint Kelsey is choosing from three different sets for each meal.

Breakfast: J   W
Lunch: B S L   B S L
Dinner: APS APS APS APS APS APS

Ask Yourself How can I represent the choices for each meal in a tree diagram?

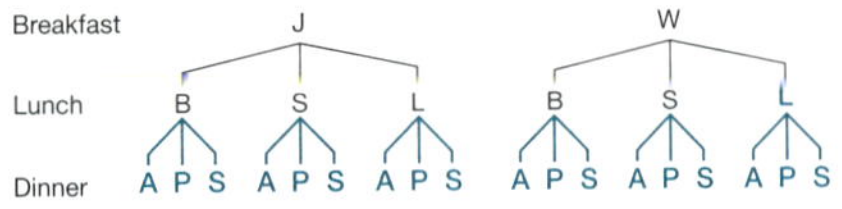

Answer There are 18 ways Kelsey can choose where to go for breakfast, lunch, and dinner.

Apply What is another question you could ask from the information given in the problem?

Sample Answer: How many choices does Kelsey have if she wants to go to Louie's Diner for lunch?

⑤ Kim's family always goes to the same sandwich shop for lunch when they are on vacation. Kim only likes ham, turkey, or cheese sandwiches. She does not want to eat the same sandwich more than two times during her three-day vacation. How many different ways are there for Kim to choose a sandwich each day?

Ask Yourself Are these combinations or permutations?

Hint Kim can have the same sandwich more than once.

HHT HTH THH HHC __HCH; CHH__
TTH THT HTT __TTC; TCT; CTT__
CCH CHC __HCC; CCT; CTC; TCC__
HTC __HCT; THC; TCH; CHT; CTH__

Answer There are 24 different ways for Kim to choose 3 sandwiches without choosing the same kind more than two times.

Relate How would your list change if Kim wanted to choose a different sandwich every day?

Sample Answer: There would only be 6 choices in my list, the ones in the last row.

190 Unit 6 Using Data and Probability

Practice

Solve the problems. Show your work.

6 Bill is packing for camp. He will be gone for one week. He packs five T-shirts: one black, one brown, one green, one red, and one blue. He packs four pairs of shorts: black, brown, tan, and blue. How many different outfits of one T-shirt and one pair of shorts can Bill make?

Some students will use the lesson strategy; however, other strategies may be used. Accept all reasonable work leading to the correct answer.

Answer Bill can make 20 different outfits.

Examine What information is given that is not needed to solve the problem?

Sample Answer: Bill will be gone for one week.

7 Sara likes four of the rides at the carnival. She likes the Ferris wheel, the bumper cars, the giant swings, and the super slide. She has enough tickets to go on three rides. She can choose to go on the same ride more than once. If order matters, how many ways are there for Sara to choose her three rides?

Some students will use the lesson strategy; however, other strategies may be used. Accept all reasonable work leading to the correct answer.

Answer There are 64 ways for Sara to choose her rides.

Analyze A student says that Sara can go on the rides in 24 ways. What mistake could the student have made?

Sample Answer: The student did not allow Sara to go on the same ride more than once.

Create Look back at the problems in the lesson. Write a new problem that can be solved by making an organized list, then solve it.

See teacher notes.

Lesson 23 Strategy Focus: Make an Organized List   191

Create

In this lesson, students write a new problem that can be solved by making an organized list or a tree diagram. If students are struggling, suggest they choose two different items and a number of different colors for each.

Accept student responses that involve a permutation or combination. Students should make it clear whether a choice can be repeated. Responses should include a correct solution using an organized list or tree diagram.

Independent Practice
Practice

Students should be encouraged to choose any strategy to solve Problems 6 and 7, though many may prefer to use *Make an Organized List*.

6 Some students may draw a tree diagram to solve the problem, while others may make an organized list. Be sure students are able to explain how they avoided omissions and repetitions.

Sample Work

Bk = Black, Br = Brown, G = Green, R = Red, Bl = Blue, T = Tan

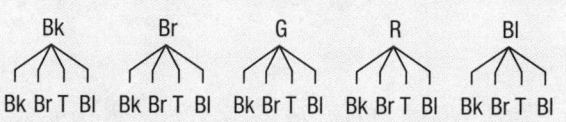

```
   Bk           Br           G            R           Bl
  /|\\         /|\\         /|\\         /|\\         /|\\
Bk Br T Bl  Bk Br T Bl  Bk Br T Bl  Bk Br T Bl  Bk Br T Bl
```

Examine (HOTS) Responses should note that the information about the length of time Bill will be away at camp is not necessary.

7 Make sure students understand that while Sara can go on a ride one, two, or three times, she cannot go on all four rides.

Sample Work

F = Ferris Wheel, B = Bumper Cars
G = Giant Swing, S = Super Slide

FFF FFB FFG FFS FBF FGF FSF

FBB FGG FSS FBG FGB FBS FSB FGS FSG

BBB BBG BBS BBF BFB BGB BSB

BFF BGG BSS BFG BGF BFS BSF BGS BSG

GGG GGF GGB GGS GFG GBG GSG

GFF GBB GSS GBF GFB GBS GSB GFS GSF

SSS SSF SSB SSG SFS SBS SGS

SFF SBB SGG SFB SBF SFG SGF SBG SGB

Analyze (HOTS) Responses should demonstrate an understanding that not allowing repeats on a ride would result in fewer choices.

Lesson 23   125

Lesson 24: Strategy Focus — Make an Organized List

Lesson Overview

Lesson Materials: calculator

Skills to Know	Outcome	Math Vocabulary	eResources www.optionspublishing.com
• List outcomes and identify favorable outcomes • Find theoretical probabilities	Students will recognize that making an organized list is an efficient way to find possible outcomes.	dependent events, independent events, probability	• Interactive Whiteboard Transparency 24 • Homework, Unit 6 Lesson 24 • Know-Find-Use Table • Problem-Solving Checklist, also available in the student worktext, page 7

Modeled Instruction

Learn

Ask questions about the problem's context to support students' comprehension of the problem.

- *How many shirts does Simon have to choose from on the first day? The second day? The third day? How do you know?*
- *How could you ask the question in the problem another way?*

As students read the problem again, pose questions to help them recognize important phrases and facts.

- *What words tell you that Simon chooses his shirt randomly?*

Use a Graphic Organizer You may wish to use this Think Aloud to demonstrate how to use a graphic organizer to identify information.

This problem is about the order Simon wears his shirts. I will reread the problem and then fill in the Know-Find-Use Table to help me decide how I might solve the problem.

In the Know column, I can write Simon has 3 shirts—one red, one white, and one blue. *I also know that* he is camping for 3 days *and* Each day he will wear a different shirt. *I can also write information about how he chooses a shirt in the Know column:* He chooses the shirt at random from his backpack and he does not put a shirt back in the backpack. *In the Find column, I can write* The probability that Simon wears a white shirt on the day after he wears a red shirt. *In the Use column, I can write the probability formula,* $P = \frac{\text{number of favorable outcomes}}{\text{total number of outcomes}}$

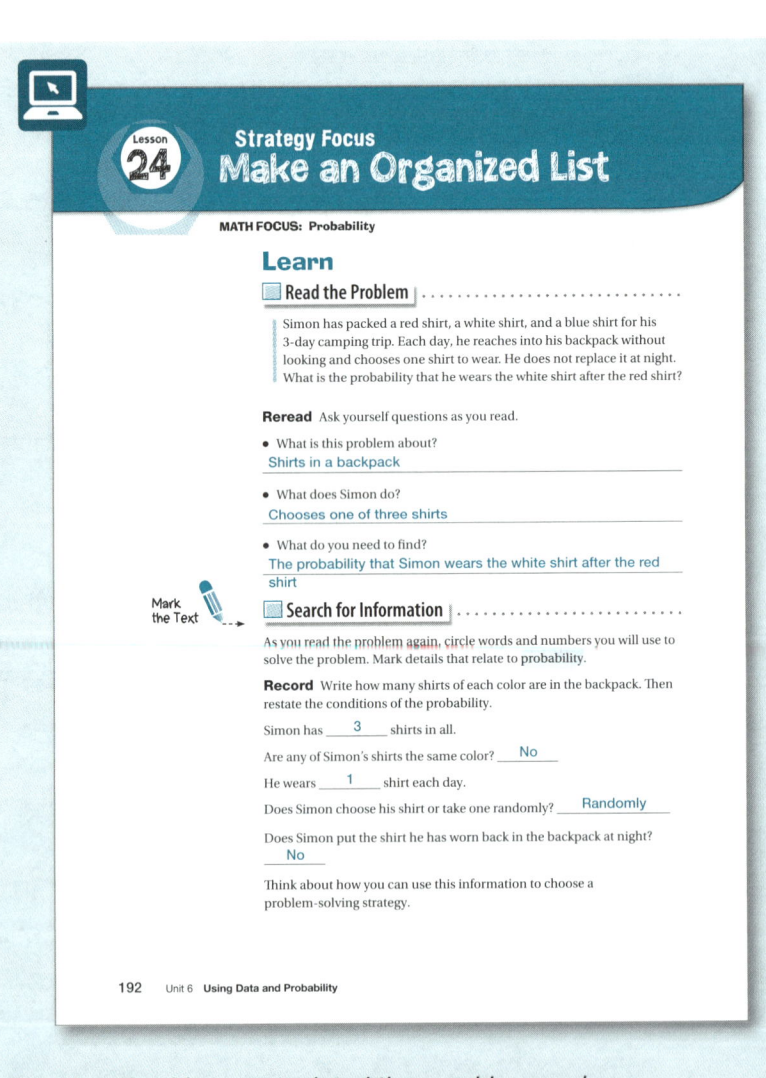

Lesson 24: Make an Organized List

MATH FOCUS: Probability

Learn

Read the Problem

Simon has packed a red shirt, a white shirt, and a blue shirt for his 3-day camping trip. Each day, he reaches into his backpack without looking and chooses one shirt to wear. He does not replace it at night. What is the probability that he wears the white shirt after the red shirt?

Reread Ask yourself questions as you read.

- What is this problem about?
  Shirts in a backpack
- What does Simon do?
  Chooses one of three shirts
- What do you need to find?
  The probability that Simon wears the white shirt after the red shirt

Search for Information

As you read the problem again, circle words and numbers you will use to solve the problem. Mark details that relate to probability.

Record Write how many shirts of each color are in the backpack. Then restate the conditions of the probability.

Simon has __3__ shirts in all.
Are any of Simon's shirts the same color? __No__
He wears __1__ shirt each day.
Does Simon choose his shirt or take one randomly? __Randomly__
Does Simon put the shirt he has worn back in the backpack at night? __No__

Think about how you can use this information to choose a problem-solving strategy.

192 Unit 6 Using Data and Probability

Now that I have completed the graphic organizer I can take what I know about the problem and what I know about probability to answer the question.

126 Unit 6

Decide What to Do

You know the number of shirts Simon has. You know he chooses a different shirt each day without looking and only wears it once.

Ask How do I find the probability that Simon wears the white shirt after the red shirt?

- I can *Make an Organized List* to show the orders that Simon can choose the shirts. Then I can use my list to find the probability of choosing the white shirt after the red shirt.

Use Your Ideas

Step 1 List the ways Simon can choose the shirts. Since he does not replace the shirt at night, these are **dependent events**.

Start by listing the ways if Simon chooses red first, then if he chooses white first, and finally, if he chooses blue first.

Red first: RWB   RBW
White first: WRB   __WBR__
Blue first: __BRW__   BWR

Use R for red, W for white, and B for blue.

Step 2 Count the ways Simon can choose white after red. The ways he can choose white after red are RWB, __RBW__, and __BRW__.

There are __3__ ways to choose white after red.

Step 3 Find the probability.

$P = \frac{\text{number of favorable outcomes}}{\text{total number of outcomes}}$

$P(\text{blue}) = \frac{\text{number of white after red}}{\text{total number of ways}}$

$= \frac{3}{6}$

So the probability that Simon wears the white shirt after the red shirt is $\frac{1}{2}$.

Review Your Work

Did you remember to simplify your answer?

Review How did making an organized list help you solve the problem?

Sample Answer: The list helped me make sure I thought of every possible way Simon could choose the shirts and made it easy to find the favorable outcomes.

193

Modeled Instruction (continued)

Help students recognize how the details they have identified can be used to determine a problem-solving strategy.

- *To write the probability of an event, what do you need to know?*
- *How can you use an organized list to help you find the number of possible outcomes and the number of favorable outcomes for this problem?*

Ask questions that encourage students to think critically about the steps in the solution process.

- *Why is it important to know that Simon does not replace the shirt at night?*

Remind students that they should always simplify their answers.

HOTS Review Students' explanations should suggest that an organized list makes it easier to efficiently and accurately find all possible outcomes.

Guided Practice

Try

Solve the problem.

① Diane's family is going on a 2-day car trip. Diane's father has the three children pick cards each day to see who gets to sit by a window. He has cards marked W, W, and M for the two window seats and the middle seat. Each day, one child after the other picks a card and does not replace it. If Diane picks first each day, what is the probability that she gets a window seat both days?

Mark the Text

Read the Problem and Search for Information

Visualize the situation in the problem.

Decide What to Do and Use Your Ideas

You can use the strategy *Make an Organized List* to help you solve the problem. These are independent events.

Ask Yourself: Does picking a W on the first day affect the probability of picking a W on the second day?

Step 1 Make a tree diagram showing the possible results of picking a card from the set of cards both days.

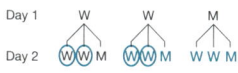

Day 1:   W   W   M
Day 2: WWM   WWM   WWM

Step 2 Use the tree diagram. Circle the outcomes that show choosing a W two times in a row.

How many possible outcomes are there? __9__

How many of those outcomes mean sitting by a window both days? __4__

So the probability of choosing W two times in a row is $\frac{4}{9}$.

Review Your Work

Check that you counted the total number of outcomes correctly.

Explain Why might a tree diagram be more helpful to you for this problem than making a list?

Sample Answer: There are two W cards. Making a list of outcomes with more than one of the same letter might be confusing.

194   Unit 6 Using Data and Probability

Try

① Prompt students to differentiate between dependent and independent events.

Help students understand the context of the problem.

- *What cards would Diane need to pick to get a window seat both days?*

Have students explain how they completed the tree diagram.

- *There are only two types of card: W and M. Why is W written two times for Day 1?*

Have students explain how they know that they have the correct total number of outcomes.

HOTS Explain Responses should note that in problems such as this where there are two W cards, tree diagrams can be a clearer way of organizing the possible outcomes.

Lesson 24   127

Scaffolded Practice
Apply

(2) Make sure students read the question carefully and notice the phrase *at least*. To help students solve this problem, ask questions like these.
- What will a favorable outcome look like in the tree diagram?
- To solve this problem, is it necessary to know whether there is replacement or not? Explain.

HOTS Identify Responses should explain that the outfit with the blue shirt and blue shorts was counted twice.

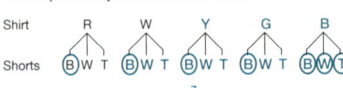

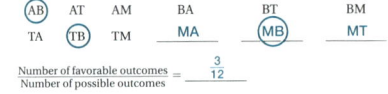

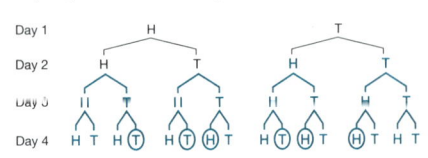

(3) Prompt students to consider how many different activities to include for each item in their list.
- If there are four activity brochures, why are only two activities included in each possible outcome?

HOTS Apply Responses should describe how to use the same organized list to find favorable outcomes for other events.

(4) Make sure students understand why *H* and *T* are used in the tree diagram given that the Quinn family could either go biking or to the beach.
- What is a possible outcome that results in the Quinn family going to the beach exactly two times? What is a possible outcome that does not result in them going to the beach exactly two times?

HOTS Formulate Responses should demonstrate students' abilities to understand how the information given in a problem can be used in other ways.

(5) Remind students that it is important to include all outcomes without repeating any.
- How will you organize your list so you do not leave out any combinations? So you do not repeat any combinations?

HOTS Analyze Explanations should show an understanding of the complement of snowboarding exactly 1 day. Students may note that the result of the subtraction described is the probability of snowboarding 0, 2, or 3 days.

128   Unit 6

Practice

Solve the problems. Show your work.

6 Kyra's family always orders one meat and one vegetable on their pizza. They randomly choose from sausage, ham, pepperoni, or chicken. Then they randomly choose from peppers, onions, olives, mushrooms, or spinach. What is the probability that the pizza toppings are ham and *either* olives *or* onions?

Some students will use the lesson strategy; however, other strategies may be used. Accept all reasonable work leading to the correct answer.

Answer The probability that the pizza toppings are ham and either olives or onions is $\frac{2}{20}$, or $\frac{1}{10}$.

Consider Does it make a difference to the problem if the family chooses the meat ingredient second instead of first? Explain.

Sample Answer: No, it does not make a difference. The combinations will still be the same.

7 Will, Andy, Lauren, and Rachel each chose an activity to do with the group while on vacation. They put their names in a hat to decide the order in which they will do the activities. The names are drawn one at a time. What is the probability that Lauren gets to do her activity immediately before Rachel?

Some students will use the lesson strategy; however, other strategies may be used. Accept all reasonable work leading to the correct answer.

Answer The probability that Lauren gets to do her activity immediately before Rachel is $\frac{6}{24}$, or $\frac{1}{4}$.

Examine Which words in the problem told you that the order "Lauren, Will, Andy, Rachel" is *not* a favorable outcome?

Sample Answer: The words *immediately before* told me that the order is not a favorable outcome.

Create Look back at the problems in the lesson. Write a new probability problem that can be solved by making an organized list. Solve your problem.
See teacher notes.

Lesson 24 Strategy Focus: Make an Organized List 197

Create

In this lesson, students write a problem about probability. They should describe a situation involving dependent or independent events. If students are struggling, suggest they write a problem like one they understood in the lesson.

Accept student responses that provide a correct solution using an organized list.

Independent Practice
Practice

Students should be encouraged to choose any strategy to solve Problems 6 and 7, though many may prefer to use *Make an Organized List*.

6 Some students may use the Fundamental Counting Principle to find the number of possible outcomes. If students use a list, be sure they select different letters to represent each topping.

Sample Work

Sausage (S), Ham (H), Pepperoni (P), Chicken (C), Peppers (p), Onions (on), Olives (ol), Mushrooms (m), Spinach (s)

Sp, Son, Sol, Sm, Ss

Hp, (Hon), (Hol), Hm, Hs

Pp, Pon, Pol, Pm, Ps

Cp, Con, Col, Cm, Cs

$P = \frac{\text{number of favorable outcomes}}{\text{total number of outcomes}}$

$= \frac{2}{20} = \frac{1}{10}$

HOTS
Consider Students' explanations should indicate that order does not matter when working with independent events.

7 Make sure students understand that drawing names one at a time implies there is no replacement after a name is drawn.

Sample Work

Will (W), Andy (A), Lauren (L), Rachel (R)

(WALR), WARL, WLAR, (WLRA), WRAL, WRLA

(AWLR), AWRL, ALWR, (ALRW), ARWL, ARLW

LWAR, LWRA, LAWR, LARW, (LRWA), (LRAW)

RWAL, RWLA, RAWL, RALW, RLWA, RLAW

$P = \frac{\text{number of favorable outcomes}}{\text{total number of outcomes}}$

$= \frac{6}{24} = \frac{1}{4}$

HOTS
Examine Students' responses should identify the phrase *immediately before* as a condition for a favorable outcome.

Lesson 24 129

UNIT 6 Review

UNIT 6 Review

In this unit, you worked with three problem-solving strategies. You can often use more than one strategy to solve a problem. So if a strategy does not seem to be working, try a different one.

Check students' work throughout. Students' choices of strategies may vary.

Problem-Solving Strategies
- ✓ Make a Graph
- ✓ Look for a Pattern
- ✓ Make an Organized List

Solve each problem. Show your work. Record the strategy you use.

1. Sebastian asked people what kind of movie they like best. The circle graph below shows his data. Sebastian remembers that 22 people chose science fiction movies as their favorite. How many people were in Sebastian's movie survey?

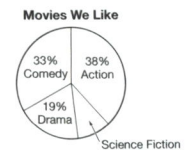

Movies We Like
- 38% Action
- 33% Comedy
- 19% Drama
- Science Fiction

Answer: There were 220 people in Sebastian's movie survey.
Strategy: *Possible Strategy:* Make a Graph

2. Alice works at a sandwich shop. The shop offers 3 kinds of meat: ham, turkey, and roast beef. The shop offers 5 toppings: pickles, lettuce, peppers, onions, and tomatoes. How many different kinds of sandwiches are there with a choice of 1 meat and 1 topping?

Answer: There are 15 different kinds of sandwiches with a choice of 1 meat and 1 topping.
Strategy: *Possible Strategy:* Make an Organized List

3. Randy times his pet mouse as it runs through a maze. Make a plot of Randy's data. Describe the correlation between the number of seconds it takes the mouse to run the maze and the number of attempts.

Attempt Number	1	2	3	4	5
Seconds	45	24	17	13	11

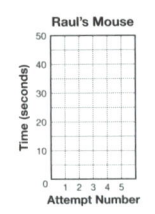

Raul's Mouse

Answer: There is a negative correlation between the attempt number and the number of seconds.
Strategy: *Possible Strategy:* Look for a Pattern

4. Kenny the Clown has 4 bow ties. There is one of each color: red, blue, green, and yellow. He has 6 pairs of clown shoes, 1 red, 1 blue, 1 green, 1 yellow, 1 orange, and 1 purple. If he chooses a tie and a pair of shoes at random, what is the probability that they will be the same color?

Answer: The probability that Kenny's tie and shoes will be the same color is $\frac{4}{24}$, or $\frac{1}{6}$.
Strategy: *Possible Strategy:* Make an Organized List

5. A town in Oregon records the amount of rainfall each month. The graph below shows the monthly totals for the town's wettest and driest years. In which months could the town expect less than 2 inches of rain to fall?

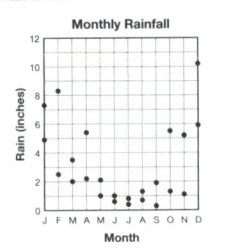

Monthly Rainfall

Answer: The town could expect less than 2 inches of rain in the months from June through September.
Strategy: *Possible Strategy:* Look for a Pattern

Explain how the graph makes it easier to solve the problem.

Sample Answer: The graph shows the monthly rainfall for the months in both the wettest year and the driest year, so the rainfall will probably be between the two values for each month.

198 Unit 6 Using Data and Probability

199

Support for Assessment

The problems on pages 198–201 reflect strategies and mathematics students used in the unit.

Although this unit focuses on three problem-solving strategies, students may use more than one strategy to solve the problems or use strategies different from the focus strategies. Provide additional support for those students who need it.

Make a Graph For Problem 1, ask students how to use the percent and number of people who watch science fiction movies to find the total number of people surveyed. For Problem 7, students may realize that they only need to compare the numbers for *doctor* and *lawyer*, since these are the only careers that were chosen by a greater percent of parents than of students.

Look for a Pattern For Problems 3 and 8, ask students how they could draw a line of best fit in the scatter plots to help them recognize a pattern. For Problem 5, have students describe the information they need to solve the problem and how they can identify it on the scatter plot.

Make an Organized List For Problems 2, 4, and 9, ask students to determine if order matters. For Problem 6, ask questions such as, *How will you organize your list? How can you use your list to find possible outcomes that have a sum of 8?* For Problem 10, make sure students understand what *at least two* implies.

You may wish to use the *Review* to assess student progress or as a comprehensive review of the unit.

Promoting 21st Century Skills

Write About It
Communication

When students explain how they used a problem-solving strategy, they have the opportunity to review the steps they used and the reasons for doing so. Responses might identify the strategy *Make an Organized List* and the steps students used to implement it.

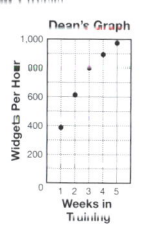

Team Project: Sales Routes
Collaboration: 3–4 students

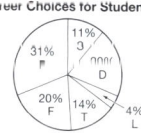

Remind students that the goal of the project is to make a group decision and work together to solve a problem based on that decision. Students should recognize that they must first choose their home city. They can then divide up the work of listing possible sales routes and calculating mileages. Finally, they should work together to organize their data and make their presentation to the class.

Ask questions that help students summarize their thinking. *How did you choose your home city? How many possible routes did you find? How did you organize your work?*

Extend the Learning
Media Literacy

If you have Internet access, navigate to sites that have online spreadsheet programs. Students may find them helpful in organizing their data and preparing their presentations for the class.

🔍 Search   spreadsheet

Unit 6   **131**

Professional Development Handbook

Contents

Problem Solving: Teacher as Coach 133
    Building on What Children Know
    Emulating Experts
    Using Your Judgment and Experience

Reading and Understanding Word Problems. 136
    Why Use Word Problems?
    Helping Students Become Active Readers
    Reading in Different Ways
    Marking the Text
    Visualizing and Representing the Problem
    Building Math Vocabulary

Meaningful Practice with Problem Solving. 140
    Good Problems Make Good Practice
    Communicating Mathematical Ideas
    Guided Practice and Differentiating Instruction

Classroom Best Practices 142
    Modeling
    What to Model
    How and When to Model

Asking Effective Questions. 145

Creating a Problem-Solving Environment 146
    Reacting to What Students Say
    Problem Solving as a Group Effort

Using Technology for Teaching and Learning 148

Bibliography 149

Problem Solving: Teacher as Coach

Building on What Children Know

Watch an infant or toddler for a few moments to see the brain at work, pondering how to reach that nearby rattle or how to make a stack of blocks. By the time children enter kindergarten, they already have five years of problem-solving experience. It makes sense to build on what children already know and what they already can do. For this reason, you should think about the job of teaching problem solving as coaching.

With guidance and practice, students can learn to apply their common sense and know-how to complex and formal mathematical situations. Trial-and-error attempts to do things, for example, grow into the more sophisticated strategy *Guess, Check, and Revise*.

The way that students approach mathematical word problems should stem naturally from their approach to problems in general. Infants trying to reach their rattles do not ask themselves: *What does Mommy want me to do?* Their approach is more along the lines of: *I know what I want. How can I get it?* That is how you want your students to approach a mathematical challenge—*What is the problem? How can I figure it out?*

Problem solving is complex, and becoming a good problem solver takes time. Like riding a bicycle, playing soccer, or making a quilt, problem solving is not a single skill that can be learned at one sitting. But with good coaching and meaningful practice, students can improve their skills and can increasingly exhibit traits found in successful problem solvers.

It makes sense to build on what children already know and what they already can do.

Emulating Experts

Good mathematical problem solvers have many strategies at their fingertips. A repertoire of tried-and-true methods is important, particularly for certain types of problems, but behaviors and attitudes play a significant role as well. Good problem solvers in any field tend to be thoughtful, persistent, reflective, flexible, confident, prepared, and willing to take risks. These are traits you want to nurture in your students.

Thoughtful Approach Students too often try anything to get through a difficult problem. Suppose a group of students is to find the area of a figure that is unfamiliar. Some might immediately start to calculate—adding or multiplying the given numbers in the hope that their result will somehow be correct. Effective problem solvers, when faced with a new problem situation, are more likely to pause and assess. They ask themselves questions to help them understand the situation.

- *What is going on in this problem?*
- *How can I figure this out?*
- *What does this remind me of?*
- *How are these quantities related?*

Persistence When stumped by a math problem, some children simply give up. Why? When students cannot find an answer right away, they assume they are just not good at math problems. Good problem solvers know that getting stuck is a natural part of the process. That awareness helps them persist.

Self-Monitoring Good problem solvers think about what they are doing. Accordingly, students should step back and examine their progress while working out a solution. Students who tend to plunge immediately into calculations will benefit from asking themselves metacognitive questions.

- *Is what I am doing helpful?*
- *If I keep going, will I be any closer to a correct answer?*
- *Is there a more efficient way to do this?*

Reflection Most students think the job is finished once they find an answer. Good problem solvers care enough about the validity of their answers to look back at their work. Students should learn to ask themselves evaluative questions.

- *Does my answer make sense?*
- *Is my reasoning valid?*
- *Does my solution answer the question asked?*
- *Do I get the same answer when I use a different approach?*

As students mature, they can learn to ask even more sophisticated questions.

- *What have I learned?*
- *Could I have solved this problem a different way?*
- *How is this problem like others I have solved before?*
- *Why did the answer come out that way?*

Confidence People who know they are going to succeed in the end are not likely to give up. Confidence breeds persistence, and success breeds confidence. Every student should experience that wonderful feeling of solving a problem that had at first seemed impossible. The problems in this book give students abundant opportunities to enjoy that accomplishment. The more that happens, the more confident students become.

Every student should experience that wonderful feeling of solving a problem that had at first seemed impossible.

Preparedness Attempting to make a pie when you have no experience in the kitchen and no knowledge of baking is unlikely to result in a tasty dessert. With solving mathematical problems, too, knowledge and practice lead to better outcomes. Thus, a fifth grader who can divide with decimals and interpret quotients is well prepared when it comes to solving an unfamiliar problem that calls for such skills.

Flexibility According to the old saying, "If at first you don't succeed, try and try again." That can be good advice for students working on mathematical problems—as long as they do not stay attached to an approach that is not working. Good problem solvers feel comfortable adjusting or replacing a method that is not serving them. To encourage this kind of flexible thinking in students, be sure they regularly experience different ways to solve the same problem.

Daring Sometimes a problem is so different from previous ones that a fresh and untried approach is called for. Good problem solvers are not afraid to try something new, knowing full well that it may not work. Many students are unwilling to take that risk because they feel they must avoid mistakes at any cost. Students need to understand that mistakes are a crucial part of learning. Making a mistake in solving a problem often helps one see a different way to approach the problem.

Using Your Judgment and Experience

Remember that you know a great deal about teaching and about problem solving. Just as students often forget to apply their common sense to math problems, some teachers forget they have a lifetime of problem-solving experience and countless hours of classroom experience they can rely on. So when you are guiding students through problems, keep in mind the following:

- You are the best person to help your students. Use your understanding of your students and your relationship with them to ask appropriate questions at appropriate times.

- Use your general experience and skills as a teacher to know what kinds of questions will help the most.

- Use your judgment of each particular situation and student to determine what kind of question to ask or whether you need to ask one.

You are the best person to help your students.

Reading and Understanding Word Problems

Why Use Word Problems?

In life outside the classroom, problems are not neatly packaged in paragraphs with well-defined questions at the end. They arise naturally, usually from a need.

- *I want to get home in time to talk with my grandparents, but I have to do some errands on the way.*
- *The game has only 5 minutes left, and we are 10 points behind.*
- *I need a notebook, a ruler, and two pens, but I only have $3 to spend.*

In an ideal world, you could wait for just the right sort of problems to come up during the school day. However, classroom practicalities do not provide that luxury. So, for expediency, teachers must use prepared word problems. Word problems can be presented at any time, and they can be designed to meet the age and ability levels of the students. That is the efficiency of word problems.

Dealing with word problems means that students have to learn how to read and understand them. Although most of the skills needed for reading word problems also apply to reading mathematics in general, some of the skills pertain only to word problems. For example, students must learn to sift through a word problem to identify what the problem is asking. That is not something we do with mathematics in real life.

... students must learn to sift through a word problem to identify what the problem is asking. That is not something we do with mathematics in real life.

Helping Students Become Active Readers

Why do so many students have trouble with problem solving? They may know the math and have good reading skills, but still misinterpret information or are unable to find a starting place for a solution. This is often because they have not yet acquired the skills needed to read a problem so that they understand the situation, know what question needs to be answered, and use that information to choose a strategy for solving the problem.

Students and adults alike often read word problems as if they are reading prose. Prose tells a story that has a beginning, a middle, and an end. A mathematical problem does not have an obvious structure; instead, there are implied relationships and conditions as well as one or more questions based on those elements. Some tools and strategies for approaching a problem can be taught; others can be invented by the solver and may be unique to that individual's thinking style. Your job is to help your students acquire these tools and strategies, along with the confidence to invent their own when needed.

Does the following problem seem familiar?

> Two trains leave at the same time, but from different cities. Train A, traveling at 60 miles per hour, leaves Northville heading toward Southville. Train B, traveling at 50 miles per hour, leaves Southville heading toward Northville. The two cities from which they leave are 330 miles apart. How long will it take for the two trains to meet? How far from each city will the trains be when they meet?

Almost everyone remembers being confounded by the context, information, and structure of the "train problem." However, when individuals are equipped with appropriate knowledge and methods, problem solving can be satisfying and empowering.

... when individuals are equipped with appropriate knowledge and methods, problem solving can be satisfying and empowering.

Reading in Different Ways

Many reading skills that students already use can be applied to reading word problems and other mathematical material, but two simple yet powerful tactics must be learned.

Reading Carefully A single sentence in a word problem can convey many ideas, and there may be charts, graphs, tables, or figures accompanying the problem. Together, the text and the visuals require interpretation and analysis. Yet, students often skim a problem, glance at the visuals, and believe they have captured all the information in one quick reading.

As a teacher and a coach, you can guide your students along a path that helps them become expert word-problem readers. First, help students recognize that problems require careful reading in order to comprehend meaning and identify details. Have students retell the problem in their own words to demonstrate that they have a good understanding of the problem's situation. Then ask them to restate the question or explain what the question is asking. Most will find that reading the problem carefully enables them to restate it effectively.

Rereading Encourage students to read a problem several times, with a different purpose in mind for each reading. The first reading should be quick; its purpose is to get a sense of what the problem is about—to see the "big picture." The second and third readings should be done slowly, with the goal of first understanding what the problem is asking and then of identifying the relevant information needed to answer the question.

Marking the Text

Encourage your students to read with a pencil in hand. It will serve them well in other content areas, too, as they take notes and discern concepts. Many students find that marking text with a highlighter is an effective way to identify key information. If students are not permitted to write on the pages that have the problems they are solving, suggest that they use a separate sheet of paper to record ideas or jot down their ideas on notes.

Reading and Understanding Word Problems   137

A typical process for marking the text involves these steps:

Step 1 Students mark the question the problem is asking—Identifying the question is the first action toward solving the problem.

Step 2 Students mark all the numbers (both numerals and words). They also mark words and phrases that contain mathematical information but are not overtly numerical (for example, terms such as *dozen*, *ton*, and *week*).

Step 3 Students identify additional words and phrases that provide clues to the steps or operations they will use to solve the problem.

Step 4 Students cross out superfluous information or unnecessary details.

After marking the text, students may be able to identify strategies, steps, and operations they can use to find a solution. In time, they will be able to prioritize steps, deciding which calculations should be performed first and which need to be done just before a final answer can be reached.

Here is an example of how marking the text could help students understand the "train problem."

> Two trains leave at the same time but from different cities. Train A, traveling at 60 miles per hour, leaves Northville heading toward Southville. Train B, traveling at 50 miles per hour, leaves Southville heading toward Northville. The two cities from which they leave are 330 miles apart. How long will it take for the two trains to meet? How far from each city will the trains be when they meet?

The second and third sentences are dense with information about the trains' speeds and locations. And notice that there are two questions to answer: *How long will it take for the two trains to meet? How far from each city will the trains be when they meet?*

Visualizing and Representing the Problem

Although the problem seems complicated, students can make it simpler to understand by drawing a diagram to represent the actions and relationships expressed in the problem. By representing the situation with a sketch, students can better see what is going on in the problem and how different quantities are related.

Northville ↓ A   60 mph

Southville ↑ B   50 mph

By representing the situation with a sketch, students can better see what is going on in the problem and how different quantities are related.

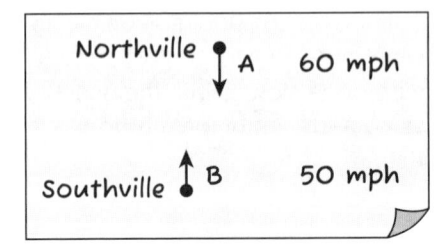

The sketch contains words, symbols, and abbreviations. Students might further cut down on writing by using the letters N and S to represent the names Northville and Southville. Such shortcuts save students time—not only when they make diagrams, but also when they write notes or make tables to represent a problem.

Once they see the big picture, students can think more clearly about how to find out where the trains will meet. They might use a strategy such as *Write an Equation, Make a Table* or *Guess, Check, and Revise.* Students might even use *Draw a Diagram* by making a more detailed drawing and marking off the motion of the two trains. Notice how drawing a diagram can be used both to help understand a problem and as a strategy for carrying out a solution.

Two Trains Traveling 330 Miles at Different Speeds

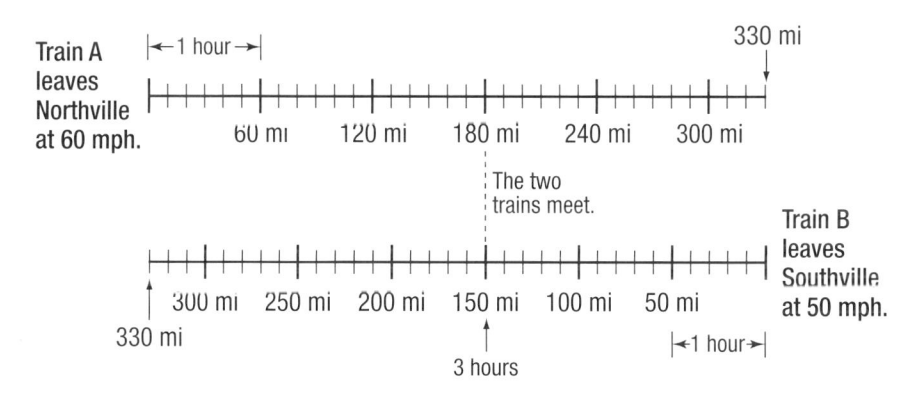

Train A leaves Northville at 60 mph. |←1 hour→| ... 330 mi
60 mi   120 mi   180 mi   240 mi   300 mi
The two trains meet.
Train B leaves Southville at 50 mph.
300 mi   250 mi   200 mi   150 mi   100 mi   50 mi
330 mi
3 hours |←1 hour→|

The train problem also offers an excellent example of how word problems require specialized reading. A written problem cannot specify every detail of reality, such as that the trains will not crash because they are on parallel tracks. When students read word problems, they must use their common sense and general knowledge to decide what assumptions to make.

Building Math Vocabulary

To understand word problems, students need to know or be able to figure out the mathematical words and phrases used. Students can develop their mathematical vocabulary by seeing and using mathematical terms in context, whether from word problems or from real-world situations in and out of the classroom. Toward that end, each unit in this book highlights and discusses a set of vocabulary words to help you make sure that students understand the terms they will use at this level.

Students can develop their mathematical vocabulary by seeing and using mathematical terms in context, whether from word problems or from real-world situations in and out of the classroom.

Reading and Understanding Word Problems

Meaningful Practice with Problem Solving

Practice makes perfect, but only when the practice is appropriate. No one becomes a proficient bicycle rider by only practicing how to get on a bike or by only memorizing the rules of the road. Improvement in bike riding requires practicing all aspects of bike riding. As children ride more often, for longer distances, and in different environments, they become better bike riders. With problem solving, too, the kind of practice that students engage in matters.

Good Problems Make Good Practice

For problem-solving practice to be meaningful, problems must help students become better problem solvers. Good problems are varied, but they share some attributes.

- Good problems cannot be solved without thinking. That is how problems and exercises differ. Exercises can be completed just by executing a standard procedure. Problems pose a challenge. They need to be understood and figured out.

- Good problems are age-appropriate. Consider this word problem: *Jen has 5 marbles. She gives 2 marbles to Sam. How many marbles does Jen have left?* This may be a problem for a child in first grade, but to a third-grader, it would be just an exercise to practice subtraction facts. Good problems challenge students without causing frustration.

- Good problems benefit students of varied ability levels. More advanced students may have an easier time getting an answer, but they still need to think. Moreover, a good problem offers opportunity for follow-up questions and explorations. *Will that result always happen? What if the amount was less? Why does that pattern occur?*

- Good problems use compelling contexts. Students are more likely to care about a solution if the problem interests them. If students are excited by the context or if they have a stake in solving the problem, they are less likely to settle for any answer just to get it over with. They are more likely to think through the problem and look back at their work.

- Good problems lend themselves to a variety of strategies. In so doing, they encourage flexible thinking and offer opportunities for communicating about mathematical methods.

- Good problems provide opportunities for building confidence. All students deserve to experience that memorable feeling when a seemingly intractable problem yields to persistent effort. Problems at an appropriate level of difficulty or complexity provide opportunities for such authentic success.

- Good problems provide practice with more than strategies. They also provide practice with skills such as estimating, finding hidden information, and ignoring extraneous information, all of which help students choose and carry out their strategies.

Good problems use compelling contexts. Students are more likely to care about a solution if the problem interests them.

Communicating Mathematical Ideas

As part of practicing problem solving, students need to practice expressing mathematical methods and reasoning. Most students find it difficult to communicate quantitative ideas effectively. They need encouragement to overcome fear and lack of experience. They also need patience, yours and theirs, because skills in writing, drawing, and speaking mathematically develop slowly over time. (See page 146, *Creating a Problem-Solving Environment,* for suggestions on promoting the development of mathematical communication skills.)

Guided Practice and Differentiating Instruction

Meaningful practice produces a good mental workout, but problems that are too difficult can lead to a feeling of failure. Providing an appropriate level of difficulty for one student is no simple task; providing it for a class with diverse ability levels is even more difficult. One way to address this issue is through guided practice, a key part of coaching.

The lessons in this book guide students through challenging problems. Each lesson provides a scaffolding system to assist students as they learn and practice the fine points of using a particular strategy. The first problem in each lesson uses questioning and modeling as students move through the process of reading the problem, deciding what question needs to be answered, and finding and using the information necessary to choose a strategy and find the solution. Each succeeding problem drops some scaffolding to leave more for the student to do independently. By the end of the lesson, students should be solving challenging problems entirely on their own.

This gradual removal of the scaffolding helps students build confidence through success. As they learn the nuts and bolts of the strategy they are practicing, they are able to handle more and more of the solution on their own. The degree to which students use the scaffolding will vary with ability level. The scaffolding provides a way for the entire class to work on the same problem so that everyone is challenged and no one is overwhelmed.

[Students] need patience, yours and theirs, because skills in writing, drawing, and speaking mathematically develop slowly over time.

Classroom Best Practices

Parents, teachers, and coaches, including problem-solving coaches, all rely on good judgment and common sense. The suggestions and tips that follow can serve as useful reminders for applying those skills in the classroom.

Modeling

Humans learn by watching and imitating. Parents know that what they do influences their children more than what they say. For students to be good problem solvers, the most important thing they need is to see good problem solving in action. The easy part of that is modeling how to solve a problem correctly from start to finish. The difficult, and more courageous, part is to demonstrate that good problem solvers travel down—and learn from—incorrect solution pathways.

For students to be good problem solvers, the most important thing they need is to see good problem solving in action.

What to Model

If students never see their teachers get stuck or stumble on a mathematical problem, how will they learn what to do when that happens to them? How can they feel comfortable that it is a part of the problem-solving experience to have difficulty or make mistakes? You can help your students become expert problem solvers if they see how you, their expert model, deal with issues like getting stuck, hitting a dead end, making an error, or not understanding the problem at first reading.

Getting Unstuck Once in a while, model being stumped for your students. Either pretend to hit an impasse or actually work on a problem that you are not sure how to solve right away. At different times, demonstrate approaches that can be used when you have no idea how to get started.

- Make sure you understand the problem. Go back and reread the problem. Check the assumptions you want to make. Ask yourself whether you are interpreting terms appropriately. Express the problem in your own words. Organize the information. Consider using a graphic organizer.

- Relate the problem to other problems you have solved. Try to think of problems that are mathematically related, but not necessarily those with the same context. A problem about finding the area of the football field would be useful to recall when needing to figure out the area of a concert stage. However, recalling that football field problem would not be very helpful when needing to figure out the cost for a family to attend a football game.

- Draw a diagram just to look at how the parts of the problem are related. It may not turn out to be an effective strategy, but it is likely to provide ideas about what strategy would be most effective.

- Identify the quantities in the problem. Visualize how they are related. Write them out, perhaps using a schematic diagram, to help you think about the connections. For example, consider this problem.

  > Maria bought some posters at $5 each and some bracelets at $3 each. She thinks she spent exactly $17 on her order. Could she be right? Explain.

  The quantities in this problem might be represented as follows.

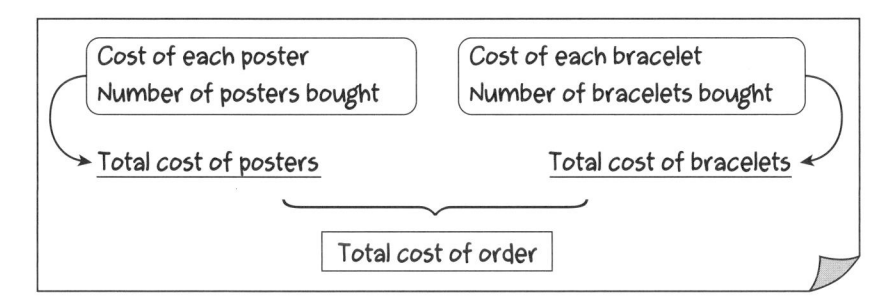

  ```
  Cost of each poster              Cost of each bracelet
  Number of posters bought         Number of bracelets bought
         ↓                                     ↓
    Total cost of posters          Total cost of bracelets
                    ⎵_____⎵
                       Total cost of order
  ```

Learning from Mistakes Students can usually fix a mistake once they see it. What is crucial to model is the acceptance of mistakes. Students need to see adults make and accept mistakes and then correct and learn from them.

What is crucial to model is the acceptance of mistakes.

Getting Past Dead Ends When modeling what to do when stuck, be sure students see that you take it in stride. Explain that you feel a bit disappointed, discouraged, or frustrated, but show that you do not give in to those feelings because you know this is part of the problem-solving process. Following are some actions that you can demonstrate when modeling what to do when a strategy leads nowhere.

- Take a break. Try saying something like this: *I do not know what to do now. I feel discouraged. I will work on a different problem and then come back to this one.* Or you could say: *I need to take a break. Let me tell you about the homework assignment. Then I will come back to finish the problem.* Explain that, at home, students might take a breather for 15 minutes or more.

- Learn from the dead-end path. Ponder aloud, *I wonder why this did not work the way I thought it would?* You might show an example where the method was valid but you had made a mistake in executing it. Another time, you might show how the strategy that is not working gives you an idea about trying another one. For example, in a problem involving the dimensions of a building for which the volume is known, a diagram may not yield the solution, but it could lead to using the strategy *Guess, Check, and Revise.*

Classroom Best Practices

Understanding a Complex Problem Any problem that a student cannot understand will seem complex. To model what to do when a problem seems incomprehensible, you can select, modify, or make up an appropriate problem and then demonstrate actions that can be taken. (See pages 136–139, *Reading and Understanding Word Problems.*) Remember to show how general reading strategies and skills can be called upon to help in decoding word problems.

Also, keep in mind that the action you demonstrate will depend on the reason the problem is not being understood in the first place. Routinely model the following:

- If you do not know what a word or phrase means, look it up, ask someone, or use the context to figure out a probable meaning.

- If the amount of information is overwhelming, go through the problem slowly, marking or otherwise recording each piece of information. Or try to locate the parts of the problem that give the big picture. For this modeling, you might use the train problem presented and discussed on pages 138 and 139.

- If a question is not clear, think about what it might mean. Then decide which meaning makes the most sense. In the solution, indicate how the question was interpreted.

How and When to Model

The more natural you can be when you demonstrate problem-solving behavior, the more effective you will be. The simplest way to be natural is to work on a problem that you actually cannot solve right away. Try that once in while if you are comfortable doing so. Use a problem you select or a problem— mathematical or otherwise—that comes up during the school day. Some of the problems in the *Use Logical Reasoning* lessons do not have readily apparent answers and would be particularly useful for modeling.

Be alert and take advantage of problems that arise naturally in class. *How should we share this? How should we decide who gets to go first? How should we choose teams?* Work with the class to solve problems like these so students experience how to work in a cooperative group setting.

When role-playing a problem solver having difficulty, the degree to which you act depends on your particular classroom. Even if students are aware that you are pretending, they can still benefit greatly from the modeling. The more fun you have with modeling, the more students will learn to have a relaxed but focused approach to solving problems.

The *Think Aloud* suggestions in the Modeling section of each lesson provide problem-specific ways to model particular techniques and behaviors. These suggestions can also give you ideas for modeling with other problems.

The more fun you have with modeling, the more students will learn to have a relaxed but focused approach to solving problems.

Asking Effective Questions

Because asking good questions is such a critical part of coaching problem solving, every problem in the program is followed by a reflective question about the problem. These questions, based on Bloom's taxonomy, help students develop the higher-order thinking skills that go hand in hand with expert problem solving.

Besides ending with a reflective question, the problems in this program also include problem-specific hints and questions for students to ask themselves. These, along with the end-of-problem questions, may inspire ideas for your own questions to ask before, while, and after students work on problems. As you formulate the questions you will ask, decide how and when to ask them, keeping in mind their different purposes.

For Guiding Good guiding questions enable students to persevere when they might otherwise give up or go too far off track. Such guidance can prevent frustration, build confidence, and help students learn new techniques. Following are some examples.

- *Have you tried …?*
- *Do you remember what you did on the problem about …?*
- *What do you know about the problem?*
- *What are you supposed to find out?*
- *How is (quantity A) related to (quantity B)?*
- *If you knew (quantity A), how would that help you solve the problem?*
- *How could you find (quantity A)?*

For Monitoring and Assessing Whenever you have the opportunity, listen to and observe closely what students say and do in response to your guidance. Your questions may uncover possible sources of difficulties such as a lack of sufficient knowledge of the mathematical content, a tendency to give up too easily, difficulty organizing, and unfamiliarity or lack of facility with problem-solving strategies or skills.

For Reflection Whenever you are about to tell a student that an answer or method is right or wrong, substitute a reflective question like one of the examples below. Reflective questions can help students become more self-reliant. Asking questions similar to the following can result in students looking back at their own answers and ideas and deciding for themselves whether they are valid.

- *How do you know?*
- *Can you tell us why that makes sense?*
- *How would you explain that to a younger brother or sister?*
- *Have you answered the question that the problem asked?*

Reflective questions can help students become more self-reliant.

Creating a Problem-Solving Environment

Students respond best to a classroom atmosphere that fosters the kind of thinking seen in expert problem solvers: thinking that is flexible, critical, creative, and independent.

Reacting to What Students Say

The way you react to your students' solutions and responses is the key to establishing a classroom environment that nurtures and enhances the development of problem-solving ability.

- Give students time to think. Research has shown that most teachers do not allow enough wait time after asking a question. Five seconds is a reasonable rule of thumb. Intervening too soon sends the wrong message about the importance of being thoughtful. Waiting too long might make students feel uncomfortable.

- Accept and encourage different approaches. Allow students to use whatever strategies they are comfortable with. At the same time, help them identify why some strategies might be more efficient than others.

- Be open to creative and unexpected responses so your students remain willing to think outside the box. With the proper environment, students may even begin to feel like there is no box.

- Be patient as students learn to express mathematical ideas. Proficiency develops over time. Provide positive reinforcement when students use a mathematical word or phrase correctly for the first time. Present models of clear and correct sentences and diagrams.

- Welcome mistakes and use them. Behind most, if not all, mistakes is an opportunity to help students learn. For example, an arithmetic error can be used to let students know that even mathematicians calculate incorrectly at times, and that when they do, they are good at catching and correcting mistakes.

Behind most, if not all, mistakes is an opportunity to help students learn.

Problem Solving as a Group Effort

Walk through a research center or any productive institution and you will find people working together to solve problems. Outside the classroom, complex problems are rarely solved by one person working in isolation. Students need to learn not only how to tackle difficult problems on their own, but also how to solve problems in cooperation with others. In addition to the problems themselves, the projects at the end of every unit represent opportunities for cooperative problem solving.

Sample Lesson Flow One common model for conducting a lesson based on cooperative problem solving is the following:

1. Present and discuss the problem. Make sure that students understand the problem before moving on.

2. Have students work in groups on the problem. Circulate and help as needed.

3. Have one or more groups, depending on time, present their results and solution methods to the class. Help students learn to ask and answer questions in a productive way.

4. Wrap up with a discussion to identify what was learned and what questions remain.

However you choose to use and present lessons, remember that you help your students most by taking on the role of coach. Encourage the best performance from your students by providing them with an abundance of opportunities to solve problems and to discuss all aspects of the problem-solving experience.

...Remember that you help your students most by taking on the role of coach.

Using Technology for Teaching and Learning

There are two things to remember when using technology in the classroom: *technology keeps changing* and *thinking is the most basic of all math skills*. Whatever technology students use, be sure they are learning to use it wisely. Help them learn to see technology as a tool, not as a crutch.

Real-World Problems

Technology enables students to work on more realistic problems, and it allows you to present problems that engage students.

Solving Real-World Problems Calculators and computers allow students to work on real-world problems that would otherwise be too complex or too tedious. Actual data are often messy, involving numbers that are cumbersome to compute using paper and pencil. Here are just a few examples of questions students can ask and answer with the help of electronic aids: *How does fuel consumption in one country compare to another? In five years, what will be the cost to make a typical movie? Which player had the best performance this season?*

Showcasing Real-World Problems Technology can be a convenient and relatively inexpensive means of presenting math problems in complete and compelling ways. The video of an actual event can impart its richness and detail far more than a word problem can. That means students get more valuable practice in deciding what information is useful, how to interpret results in light of the nuances of the given situation, and what questions remain to be asked.

Interactive Whiteboard-Ready Transparencies

Strategies for Success: Mathematics Problem Solving provides the first two pages of every lesson (*Learn About It* in grades 3–5; *Learn* in grades 6–8) as interactive whiteboard-ready transparencies. This widely accepted classroom technology is an effective teaching tool for enhancing classroom conversation, engaging students, and motivating collaboration. As you use this tool, take advantage of opportunities to bring in outside resources such as online manipulatives for use in lessons that focus on number operations, patterns, or geometry.

Bibliography

Beck, I., McKeown, M. G., and Kucan, L. (2002). *Bringing Words to Life: Robust Vocabulary Instruction*. NY: Guilford Press.

Beck, I., McKeown, M. G. and Kucan, L. (2008). *Creating Robust Vocabulary: Frequently Asked Questions and Extended Examples (Solving Problems in the Teaching of Literacy)*. NY: Guilford Press.

Beed, P. L., Hawkins, E. M., and Roller, C. M. (1991). Moving Learners Toward Independence: The Power of Scaffolded Instruction. *Reading Teacher*, (44) 9. pp. 648–655.

Bransford, J., Brown, A., and Cocking, R. (1999). *How People Learn: Brain, Mind, Experience, and School*. National Research Council, Washington, D.C.: National Academy of Sciences.

Burns, Marilyn (1995). *Writing in Math Class: A Resource for Grades 2–8*. Sausalito, CA: Math Solutions.

Carpenter, T., Fennema, E., Franke, M., Levi, L., and Empson, S. (1999). *Children's Mathematics: Cognitively Guided Instruction*. Portsmouth NH: Heinemann.

Carpenter, T., Hiebert, J., Fennema, E., and Fuson, K. (1997). *Making Sense: Teaching and Learning Mathematics with Understanding*. Portsmouth NH: Heinemann.

Charles, R., Lester, F., and O'Daffer, P. (1987). *How to Evaluate Progress in Problem Solving*. Reston, VA: NCTM.

Childress, R. and Pauley, R. (2006). *Closing the Achievement Gap, Best Practices in Teaching Mathematics*. Charlestown, WV: The Education Alliance.

Ellis, E. S., and Larkin, M. J. (1998). Strategic Instruction for Adolescents with Learning Disabilities. In B. Y. L. Wong (Ed.), *Learning About Learning Disabilities* (2nd ed.). San Diego, CA: Academic Press.

Forehand, M. (2005). Bloom's Taxonomy: Original and Revised, in M. Orey (Ed.), *Emerging Perspectives on Learning, Teaching, and Technology*.

Hiebert, J. and Carpenter, T. P. (1992). Learning and Teaching with Understanding. In D. Grouws (Ed.), *Handbook for Research on Mathematics Teaching and Learning*. New York: MacMillan.

Hogan, K. and Pressley, M. (Eds.) (1997). *Scaffolding Student Learning: Instructional Approaches and Issues*. Cambridge, MA. Brookline Books.

Kame'enui, E. J., Carnine, D. W., Dixon, R. C., Simmons, D. C. and Coyne, M. D. (2002). *Effective Teaching Strategies That Accommodate Diverse Learners* (2nd ed.). Upper Saddle River, NJ: Merrill Prentice Hall.

Kane, R. B., Byrne, M. A., and Hater, M. A. (1974). *Helping Children Read Mathematics* NY: American Book Company.

Labinowicz, E. (1980). *The Piaget Primer: Thinking, Learning, Teaching*. Menlo Park, CA: Addison-Wesley.

Liping, Ma (1999). *Knowing and Teaching Elementary Mathematics: Teachers' Understanding of Fundamental Mathematics in China and the United States*. Mahwah, NJ: Lawrence Erlbaum Associates, Inc.

Marzano, R. and Pickering, D. (2005). *Building Academic Vocabulary: Teacher's Manual*. Alexandria, VA: Association for Supervision and Curriculum Development.

Montague, Marjorie (2005). *Math Problem Solving for Primary Elementary Students with Disabilities*. Washington, D.C.: The Access Center: Improving Outcomes for All Students K-8.

Moschkovich, J. (1999). Supporting the Participation of English Language Learners in Mathematical Discussions. *For the Learning of Mathematics*, (19) 1.

Murray, M. (2004). *Teaching Mathematics Vocabulary in Context: Windows, Doors, and Secret Passageways*. Portsmouth, NH: Heinemann.

National Council of Teachers of Mathematics (2006). *Curriculum Focal Points for Prekindergarten through Grade 8 Mathematics*. Reston, VA: NCTM.

National Council of Teachers of Mathematics (2000). *NCTM Principles and Standards for School Mathematics*. Reston, VA: NCTM.

National Research Council, (2001). *Adding It Up: Helping Children Learn Mathematics.* Kilpatrick, J., Swafford, J., and Findell, B. (Eds), Mathematics Learning Study Committee, Center for Education. Washington, D.C.: National Academies Press.

O'Connell, Susan (2005). *Now I Get It: Strategies for Building Confident and Competent Mathematicians, K-6.* Portsmouth, NH: Heinemann.

O'Connor, C., Anderson, N., Chapin, S. and Gordon, T. (2002). *Classroom Discussions: Using Math Talk to Help Students Learn, Grades 1–6*. Sausalito, CA: Math Solutions.

O'Daffer, P., Charles, R., Cooney, T., Dossey, J. and Schielack, J. (2008). (4th Ed.) *Mathematics for Elementary School Teachers*. Menlo Park, CA: Addison-Wesley.

Pyke, C. (2003). The Use of Symbols, Words, and Diagrams as Indicators of Mathematical Cognition: A Causal Model. *Journal of Research in Mathematics Education,* (34) 5. pp. 406–432.

Roh, Kyeong Ha (2003). *Problem-Based Learning in Mathematics*. Columbus, OH: ERIC Clearinghouse for Science, Mathematics, and Environmental Education.

Rosenshine, B. and Meister, C. (1992). The Use of Scaffolds for Teaching Higher-Level Cognitive Strategies. *Educational Leadership*, (49) 7. pp. 26–33.

Stigler, J. and Hiebert, J. (1999). *The Teaching Gap: Best Ideas from the World's Teachers for Improving Education in the Classroom*. NY: Free Press.

Sullivan, P. and Lilburn, P. (2002). *Good Questions for Math Teaching: Why Ask Them and What to Ask, K–6*. Sausalito, CA: Math Solutions.

Templeton, S., Bear, D., Invernizzi, M., and Johnston, F. (2009). *Vocabulary Their Way: Word Study with Middle and Secondary Students*. Englewood Cliffs, NJ: Prentice Hall.

University of the State of New York (May 2006), *Instructional Recommendations for Elementary and Intermediate Mathematics*, Albany, NY: State Education Department.

Van de Walle, J. (2006). (6th Ed.) *Elementary and Middle School Mathematics: Teaching Developmentally*. Boston, MA: Pearson.

Wisconsin Center for Education Research, *A Guide to Help Teachers Understand Cognitively Guided Instruction,* Madison, WI.

Appendix: eResources Contents

Problem-Solving Toolkit
Problem-Solving Checklist

Unit and Lesson Support

Unit 1
Unit 1 Home/School Connection Letter
Unit 1 Home/School Connection Letter (Spanish)
Interactive Whiteboard Transparency 1
Homework: Lesson 1
Homework: Lesson 1 Answer Key
Interactive Whiteboard Transparency 2
Homework: Lesson 2
Homework: Lesson 2 Answer Key
Interactive Whiteboard Transparency 3
Homework: Lesson 3
Homework: Lesson 3 Answer Key
Interactive Whiteboard Transparency 4
Homework: Lesson 4
Homework: Lesson 4 Answer Key

Unit 2
Unit 2 Home/School Connection Letter
Unit 2 Home/School Connection Letter (Spanish)
Interactive Whiteboard Transparency 5
Homework: Lesson 5
Homework: Lesson 5 Answer Key
Interactive Whiteboard Transparency 6
Homework: Lesson 6
Homework: Lesson 6 Answer Key
Interactive Whiteboard Transparency 7
Homework: Lesson 7
Homework: Lesson 7 Answer Key
Interactive Whiteboard Transparency 8
Homework: Lesson 8
Homework: Lesson 8 Answer Key

Unit 3
Unit 3 Home/School Connection Letter
Unit 3 Home/School Connection Letter (Spanish)
Interactive Whiteboard Transparency 9
Homework: Lesson 9
Homework: Lesson 9 Answer Key
Interactive Whiteboard Transparency 10
Homework: Lesson 10
Homework: Lesson 10 Answer Key
Interactive Whiteboard Transparency 11
Homework: Lesson 11
Homework: Lesson 11 Answer Key
Interactive Whiteboard Transparency 12
Homework: Lesson 12
Homework: Lesson 12 Answer Key

Unit 4
Unit 4 Home/School Connection Letter
Unit 4 Home/School Connection Letter (Spanish)
Interactive Whiteboard Transparency 13
Homework: Lesson 13
Homework: Lesson 13 Answer Key
Interactive Whiteboard Transparency 14
Homework: Lesson 14
Homework: Lesson 14 Answer Key
Interactive Whiteboard Transparency 15
Homework: Lesson 15
Homework: Lesson 15 Answer Key
Interactive Whiteboard Transparency 16
Homework: Lesson 16
Homework: Lesson 16 Answer Key

Unit 5

    Unit 5 Home/School Connection Letter
    Unit 5 Home/School Connection Letter (Spanish)
    Interactive Whiteboard Transparency 17
    Homework: Lesson 17
    Homework: Lesson 17 Answer Key
    Interactive Whiteboard Transparency 18
    Homework: Lesson 18
    Homework: Lesson 18 Answer Key
    Interactive Whiteboard Transparency 19
    Homework: Lesson 19
    Homework: Lesson 19 Answer Key
    Interactive Whiteboard Transparency 20
    Homework: Lesson 20
    Homework: Lesson 20 Answer Key

Unit 6

    Unit 6 Home/School Connection Letter
    Unit 6 Home/School Connection Letter (Spanish)
    Interactive Whiteboard Transparency 21
    Homework: Lesson 21
    Homework: Lesson 21 Answer Key
    Interactive Whiteboard Transparency 22
    Homework: Lesson 22
    Homework: Lesson 22 Answer Key
    Interactive Whiteboard Transparency 23
    Homework: Lesson 23
    Homework: Lesson 23 Answer Key
    Interactive Whiteboard Transparency 24
    Homework: Lesson 24
    Homework: Lesson 24 Answer Key

Additional Reproducible Resources

Graphic Organizers

    Problem-Solving Diagram
    Know-Find Table
    Know-Find-Use Table
    Word Circle 1
    Word Circle 2
    Word Diagram
    Word Map
    Word Web 1
    Word Web 2
    Word Web 3
    Word Web 4

Math Tools

    Grid Paper — 1 cm
    Grid Paper — 5 mm
    First Quadrant Grids
    Coordinate Grid 1
    Coordinate Grid 2
    Coordinate Grids
    10×10 Grids